Aspects of Geography

General Editors: Keith Clayton and J.H. Johnson

Glaciation

Cuchlaine King

Macmillan Education
London and Basingstoke

First published 1984

Published by
MACMILLAN EDUCATION LTD
Houndmills Basingstoke Hampshire RG21 2XS
and London
Associated companies throughout the world

Typeset by Oxprint Ltd, Oxford

Printed in Hong Kong
ISBN 0 333 37347 2

Acknowledgements

Figures 4, 5 and 7 are taken from *Glacial systems, an approach to glaciers and their environments* by J. T. Andrews. © 1975 by Wadsworth Publishing Company, Inc. Reprinted by permission of Wadsworth Publishing Company, Belmont, California 94002, USA.

The publishers have made every effort to trace copyright holders, but if they have inadvertently overlooked any they will be pleased to make the necessary arrangements at the first opportunity.

Contents

Preface

In recent years, geography has been changing with great speed. It is not primarily that the basic facts of geographical distributions are themselves changing, although this has happened. It is far more that geographers have come to think differently about the significance of geographical distributions, about how to study them, and about what topics are worthy objects of geographical investigation.

Nobody can remain in close touch with the expanding frontier of geographical knowledge at all its points. New developments in the subject appear as papers in learned journals, but these are hard to track down and, even when found, they are difficult for the non-specialist to assess. Nor can all new developments be taken up quickly by the standard textbooks, which must necessarily go some years between revisions. As a result, *Aspects of Geography* has been organised as a series of concise reports by writers who are in contact with a particular sector of the subject's development. Although the series is particularly aimed at the inquiring A-level student and his teacher, it is hoped that the series will also be useful to college and university students as an introduction to the various specialist fields that will be covered.

It is sometimes difficult to realise that as little as 15 000 years ago much of the British Isles was covered by an ice sheet which rose to a thickness of almost 2 km over Scotland. Yet once we are alerted to the signs of this past glaciation – the U-shaped troughs, the rock basin lakes, the extensive sheets of till, and such landforms as drumlins and eskers – the signs are clear enough. Indeed, much of the most exciting mountain scenery of the British Isles is the result of glaciation, while, but for glacial deposits, quite large areas of eastern England would be below sea level. Even the steep chalk escarpments of the North Downs or the Chilterns are missing from East Anglia as a result of glaciation. The Upper Dee now flows to the Severn, the Thames has been diverted from the Vale of St. Albans to the southern edge of the London Basin.

Cuchlaine King explains how ice moves, erodes and deposits, and how the climatic changes of the Ice Age led to a repeated series of glacial periods. She explains how the tell-tale forms resulting from glaciation may be recognised in the landscape, and emphasises that in geological terms we are the inhabitants of an interglacial period which, if past events are a secure guide, will sooner or later be terminated by a return to glacial conditions. In this way her account ties in well with another book in this series, *Climatic Change* by R.G. Barry.

KEITH CLAYTON
J.H. JOHNSON

Ice ages

Introduction

Glaciation has played an important part in giving the earth's surface its present character. Much of the land has been directly affected by ice, but all the earth has been indirectly affected in many ways. An ice age is usually a prolonged period of tens of millions of years, during which there is ice present somewhere upon the earth. At present we are living in an ice age, although much of geological time has been without ice. However, the extent of the ice cover varies, giving glacial and warmer, interglacial episodes within the ice age.

There appear to have been at least seven ice ages before the present one, and possibly more. They have been spaced very roughly 150 million years apart, four having occurred in the pre-Cambrian more than 600 million years ago. The most recent ones occurred in the late Ordovician, in the Permo-Carboniferous, about 300 million years ago, and arguably in the Jurassic about 150 million years ago. The present Ice Age started during the Tertiary about 10 to 20 million years ago, although ice probably did not affect Britain directly until much more recently. The periodicity in the time spacing of major ice ages gives some clues as to their causes, which will be considered later in this chapter. First, however, a brief introduction to the distribution of ice now and at the maximum of the last major ice advance will be given. This was not the absolute maximum in many areas. A discussion of the time sequence, or chronology, of the very significant fluctuations within the Ice Age follows. The rest of the book will be devoted to a discussion of ice and its behaviour, and its effects in the uplands and lowlands. The glaciation of Britain will then be considered briefly.

Present ice cover

Nearly all the ice on the earth today is in large ice sheets in Antarctica and Greenland. The former covers an estimated 12.5 million km², and the latter 1.7 million km², out of a total world coverage of very nearly 15 million km². Table 1 gives the present areas covered by ice, and those of the last major advance about 20 000 years ago. Ice masses are found in all latitudes, and their distribution depends on a number of factors. They will form where more snow falls than melts during a year; thus, they are mostly found in areas of fairly high snowfall and low temperatures.

Region	Present area (km²)	Maximum Quaternary area (km²)
Antarctica	12 535 000	13 800 000
Greenland	1 726 400	2 295 300
Laurentide complex	147 248	13 386 964
Fennoscandia	3 810	6 666 708*
Rocky Mtns and S. Alaska	76 880	2 610 127
Asia	115 021	3 951 000
Alps, Europe	3 600	c. 37 000
South America	26 500	870 000
Australasia	1 015	30 000
Total	14 898 320	44 383 436

Table 1 Area of present and maximum ice cover in different regions (R.F. Flint, 1971)
The total includes other small complexes not included in the areas listed
* includes ice in UK

Ice cover at the maximum of the last glaciation

The area covered by ice during the maximum extent of the last glaciation, which was about 20 000 years ago, was about three times the area now covered by ice (table 1). The major ice sheets in Antarctica and Greenland were not very much larger in area then, because their size is limited by that of the land masses on which they lie. This is because thick ice cannot develop over the sea. The Antarctic ice sheet is estimated to have had an area over 10 per cent larger, while the Greenland ice sheet was over 35 per cent larger. The greatly increased world extent of ice cover was due to the growth of the two large, land-based ice sheets of the northern hemisphere, the Scandinavian ice sheet and the Laurentide ice sheet in North America. The former was about half the size of the latter, which was bigger than the Antarctic ice sheet is now. The Asian ice masses were also much larger, and the Alpine ice sheet was ten times larger, while extensive areas of western North America and South America were ice covered. The total ice cover at this time is estimated as over 44 million km².

Sea ice

One of the main reasons for the great difference in ice cover in glacial and interglacial conditions is the distribution of land and sea. Sea ice is mainly formed by the freezing of sea water when the temperature falls low enough, depending on the salt content, or salinity, of the sea water: the more saline the water, the lower the temperature necessary to freeze it. Normal ocean salinity is about 35 parts per thousand throughout, at which value the freezing point of water is at a temparature of −1.9° C. Once formed the pack ice drifts with the currents, and usually it drifts to warmer waters where it melts before it is 10 years old. At this age its thickness is about 4 m. Greater thicknesses occur where tide, winds or currents force the ice floes together to form pressure ridges. Pack ice has a strong seasonal distribution. Around Antarctica about 18 to 20 million km^2 of ocean are ice covered at the maximum extent, the value falls to a minimum of about 2.5 million km^2 in March, the southern hemisphere autumn. The largest area of pack ice in the northern hemisphere is in the Arctic ocean. Its area is about 11.7 million km^2 in winter, falling to not less than 8.2 million km^2 in summer. The smaller variation is due to the landlocked nature of the Arctic ocean.

Sea ice does play an important part in climatic fluctuations because it increases cooling of the earth. This is because much of the incoming solar radiation is reflected from the ice surface. Ice has a high *albedo*, or reflectivity. This albedo of ice is much higher than that of a water surface. Heat can also penetrate through a much greater depth of water than of either snow and ice or exposed land surfaces, so that more heat can be absorbed by water.

Land ice

Ice that forms on land can accumulate to very great thicknesses, and can spread far from its source under favourable conditions, to form massive and long-lasting or stable ice sheets, such as that covering Antarctica. It is only where large land areas occur in fairly high latitudes, as in the northern hemisphere, that ice sheets can form. If the land areas extend too close to the limit of ice formation, however, the ice sheet will be unstable. The Laurentide and Scandinavian ice sheets illustrate these unstable ice masses. Their fluctuations have been largely responsible for the glacial and interglacial phases.

Northern hemisphere land-based ice sheets

The major contrast between glacial and interglacial periods is the presence during the glacials of the large land-based ice sheets of the northern hemisphere. The Scandinavian ice sheet started to form on the high plateau of Norway, where its last remnant, the Jostedalsbreen, is now situated. This plateau, which was raised up in late Tertiary times, lies athwart the snow-bearing westerly winds. As R.F. Flint (1971) suggested, it is probable that snow first started to accumulate on this upland, forming vigorous glaciers that flowed down to the open Atlantic ocean to the west. Glaciers flowing eastwards had no free outlet to the sea, but spread over extensive low-lying land, where the ice thickness could increase to at least 3000 m. As this ice mass built up, so the ice parting was transferred east, as indicated on figure 1. This idea is supported by the occurrence of Swedish erratics in Norway, indicating a westward movement of the ice. An erratic is a rock carried by ice into an area of different rock type.

At its maximum the Scandinavian ice sheet joined the British ice in the North Sea basin, and at times impinged on the coast of north-east England. The North Sea basin would have been dry land then, owing to the fall of sea level as water was withdrawn to form ice when the climate cooled. The major spread of the ice sheet was,

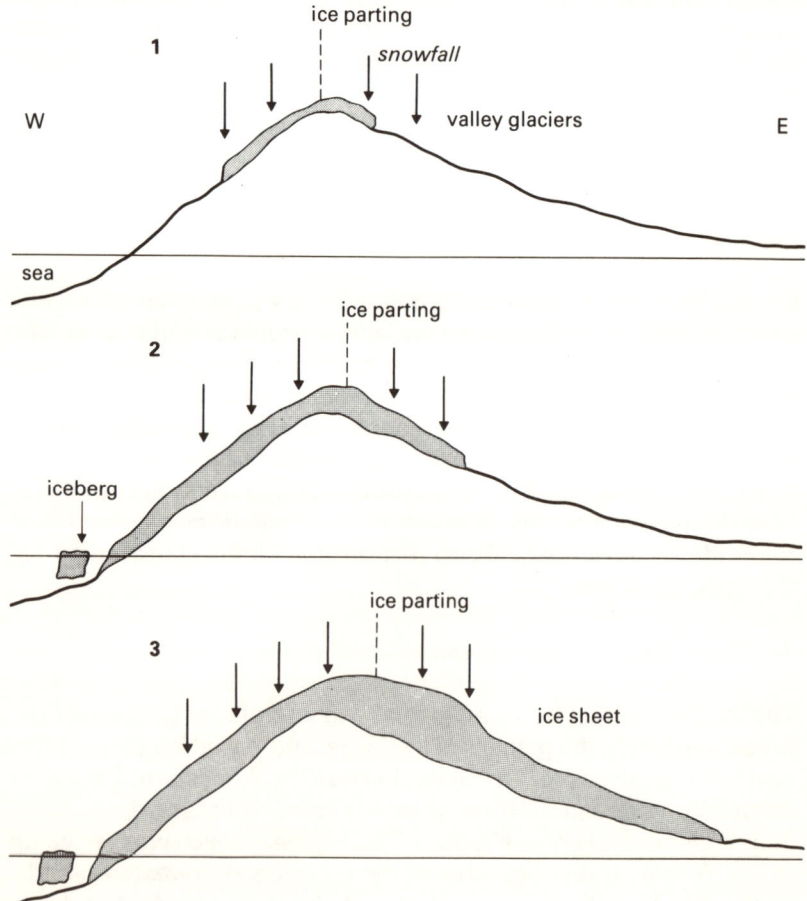

Figure 1 Diagrammatic profiles to illustrate the possible development of the Scandinavian ice sheet, following the ideas of R.F. Flint

however, to the south and east, to cover the lowlands of Poland, Germany, all Denmark and much of Holland, as shown on the map, figure 2. At its maximum in the Saalian glaciation (see table 3), the ice spread in two large lobes into the Don and Dnepr valleys. The ice sheet also probably linked with one over western Siberia.

The Laurentide ice sheet at its maximum was nearly the same size as the Antarctic ice sheet. As the climate deteriorated, small cirque glaciers (see p. 19) would grow along the uplifted margin of the Canadian shield. There were three main centres: the uplands of Baffin Island, the highlands of Labrador, and the Keewatin area to the west of Hudson Bay. Some of the Arctic Islands, including Ellesmere and Axel Heiberg, also probably accumulated some ice. Apart from the highlands there are also extensive plateaux at about

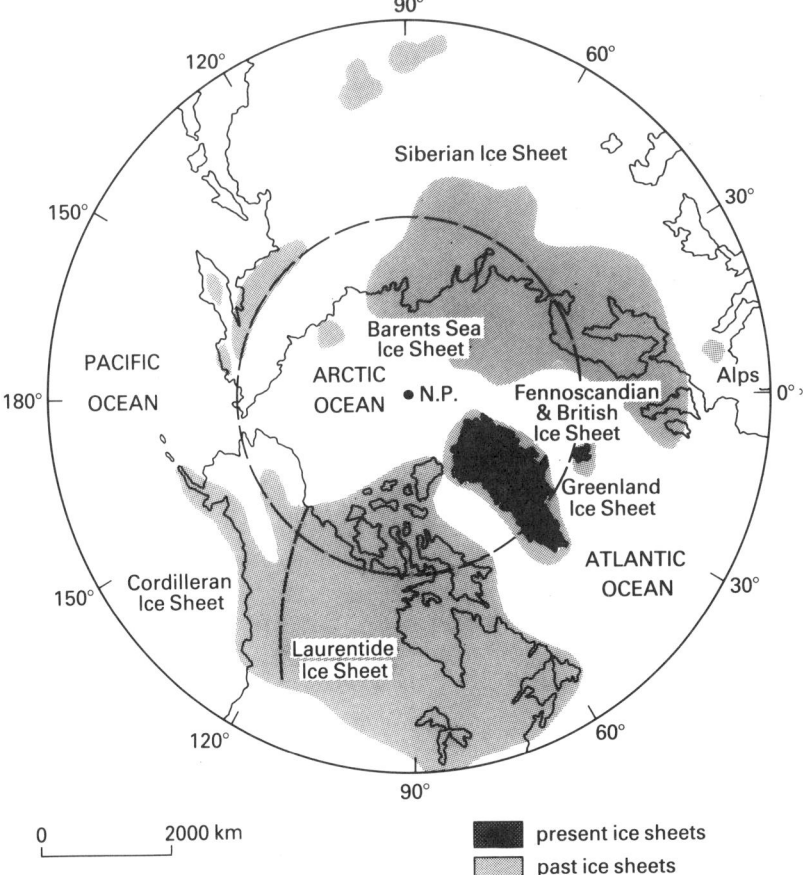

Figure 2 Map to show the distribution of northern hemisphere ice sheets at the maximum glaciation and at present (after J.T. Andrews, 1975)

5

500 to 700 m. As the snow line fell these plateau areas would probably undergo very rapid glacier development: this theory has been referred to as 'instantaneous glacierization' by J.D. Ives *et al.* (1975). The growing snow patches would rapidly join to form an extensive ice sheet that spread very far to the west and south, reaching New York state on the east and linking at some stages probably with the Cordilleran ice sheet of western Canada in the west.

Causes and chronology of ice ages

The apparent periodicity of ice ages in geological time has already been noted. This fact gives some clues as to the possible causes of major ice ages. There is probably some connection between the ice ages and the main periods of mountain building. This seems reasonable, as ice is more likely to form in areas of high relief. The high land must, however, be in suitable latitudes where climatic conditions allow an extensive ice cover to develop. Theories of continental drift, therefore, and the more recent explanation in terms of plate tectonics, also help to explain the initiation of ice ages. There are also important fluctuations within an ice age, and some cause external to the earth itself is responsible for these variations between glacial and interglacial periods.

One of the best researched of the older ice ages was that in the Permo-Carboniferous period. Evidence for ice at this time is found mainly in widely separated parts of the southern hemisphere, including South Africa, South America, India and Australia, all of which are now in relatively low latitudes. This distribution gave Alfred Wegener important evidence for his theory of continental drift. Wegener was a climatologist, and in seeking an explanation for the distribution of evidence of glaciation and cold type flora, he was struck by the apparent jig-saw fit of the African and South American coastlines. He therefore suggested that the southern lands had at one time formed one single land mass, which he called Pangaea. He thought that this large land area was centred close to the south pole at this time, and that it has since broken up and moved to lower latitudes. Modern work on plate tectonics has confirmed and supported his theory, which is now generally accepted.

The Tertiary was a time of active mountain building, when the present high mountains were mostly uplifted. These include the Rocky Mountains and Andes along the western edge of the Americas, the Alps of Europe and the Himalayas in Asia, and the Southern Alps of New Zealand. Uplift to the extent of 2000 m in less than 10 million years occurred in the European Alps, and the Himalayas rose about 3000 m in the same time. There was also renewed uplift of older structures in Scandinavia and Britain as well as other parts of North America. Similar events affected Antarctica.

It is significant that the world climate started to cool about 50 million years ago as Antarctica drifted south from Australia to a high latitude area. The elevation of the Transantarctic Mountains at about the same time may have been responsible for the initiation of the Ice Age, which probably started in Antarctica at about this time. Ice did not start to form glaciers and ice sheets in the northern hemisphere until considerably later. There is evidence of ice in Alaska about 10 million years ago in the form of glaciers. By this time the ice sheet in Antarctica was about half its present size, and it has never melted since. It may even have been larger than now about 5 million years ago. The major northern hemisphere ice sheets probably first formed at the beginning of the Pleistocene period, about 2 to 3 million years ago. Table 2 illustrates these events.

One possible cause for the initiation of the northern ice sheets has been suggested by A.T. Wilson. He thought greater snowfall in Antarctica would occur during warmer phases. The higher temperature in the south would allow a greater moisture content of the air and hence greater precipitation as snow. The greater snowfall would cause a thickening of the ice sheet which would raise the subglacial temperature to the pressure melting point. The ice could then flow more easily and a massive advance would result. The greater extent of the southern ice would, he suggested, alter the general circulation of the atmosphere and cool the global climate enough to start glaciation in the northern hemisphere. The higher albedo of the larger area of ice-covered ocean would also help to lower world temperatures.

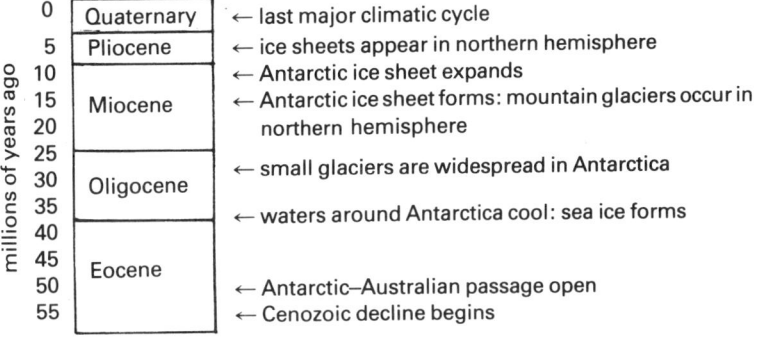

millions of years ago		
0	Quaternary	← last major climatic cycle
5	Pliocene	← ice sheets appear in northern hemisphere
10		← Antarctic ice sheet expands
15	Miocene	← Antarctic ice sheet forms: mountain glaciers occur in
20		northern hemisphere
25		
30	Oligocene	← small glaciers are widespread in Antarctica
35		← waters around Antarctica cool: sea ice forms
40		
45	Eocene	
50		← Antarctic–Australian passage open
55		← Cenozoic decline begins

Table 2   The development of the present ice age in the last 55 million years

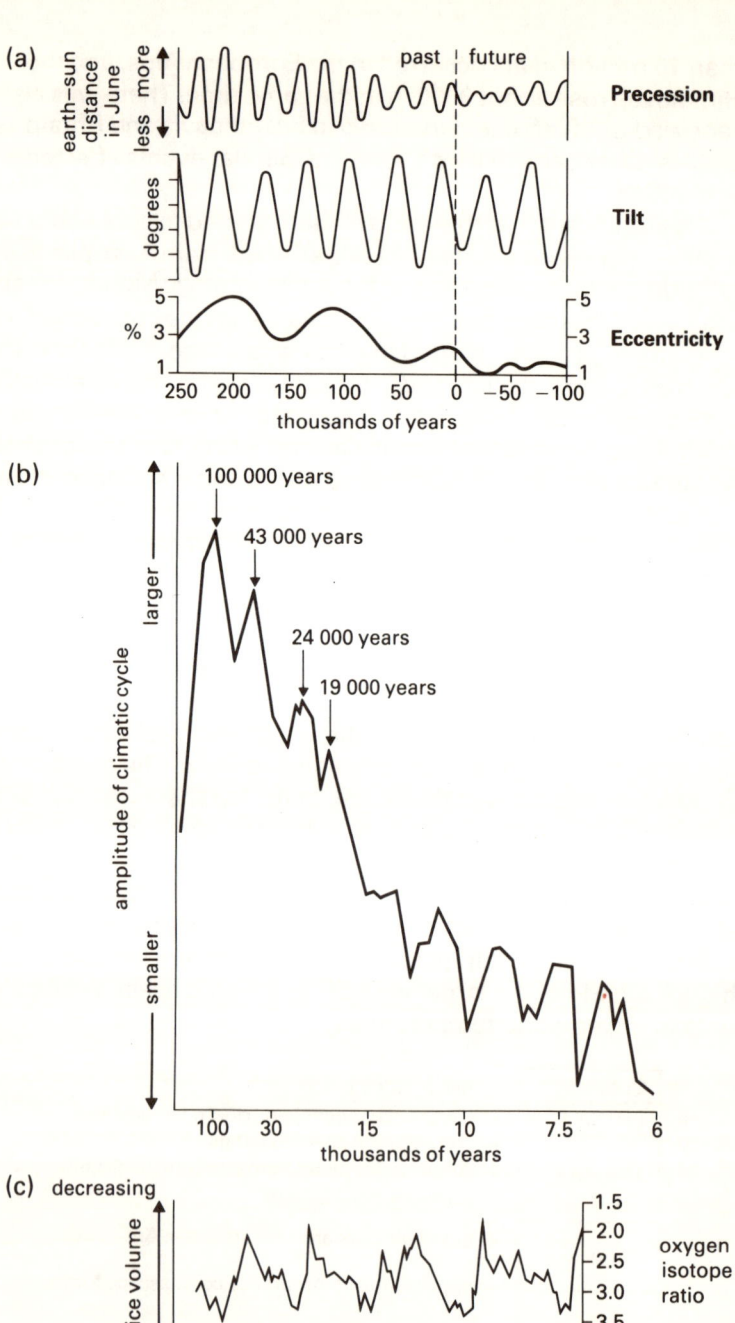

(a)

earth–sun distance in June — less / more

Precession

past | future

degrees — Tilt

% — Eccentricity

5 / 3 / 1

5 / 3 / 1

250 200 150 100 50 0 −50 −100

thousands of years

(b)

amplitude of climatic cycle — larger / smaller

100 000 years

43 000 years

24 000 years

19 000 years

100 30 15 10 7.5 6

thousands of years

(c) decreasing

ice volume — increasing

1.5
2.0
2.5
3.0
3.5
4.0

oxygen isotope ratio

500 400 300 200 100 present

thousands of years ago

There have been many phases of glacier advance with intervening periods of retreat, giving glacial and interglacial periods. Within each glacial there are also colder stadials and warmer interstadials, and even shorter climatic fluctuations occur. There is a marked periodicity in these climatic changes, which was first pointed out by Croll in the mid-nineteenth century. The periodicities can be related to the movements of the earth relative to the sun, and so to the intensities of solar radiation that reach the earth.

The astronomical cycles were elaborated by Milankovich in the early decades of the twentieth century. He established three cycles. One of these has a period of 22 000 years and is called the precession cycle. It is related to the varying distance between the earth and the sun. The second cycle is the variation in the tilt of the earth's path round the sun and it has a period of 41 000 years. The path of the earth round the sun is called the ecliptic, and is now tilted at an angle of 23½ degrees to the equator, but it varies slightly over time. The third cycle has a period of 91 800 years and is due to changes in the eccentricity of the earth's orbit around the sun. These cycles, which are shown in figure 3(a), affect different latitudes differently, and they can be superimposed to form a complex sequence of variations in solar energy receipt at various latitudes. The result of superimposing the three cycles is to give major cycles of approximately 100 000 years, with lesser peaks at 43 000 years, 24 000 years and 19 000 years, as shown in figure 3(b). Figure 3(c) illustrates the climatic fluctuations over time, derived from oxygen isotope ratio data. The dominance of the 100 000 year cycle is clearly shown, and the lesser fluctuations are also apparent.

A great deal of research in a wide range of fields has been carried out to substantiate the view that these astronomical cycles are responsible for the periodicity of climatic change and glacial fluctuations during the last million or so years. Some of the best evidence for this theory of the ice age fluctuations has been derived from the study of deep sea cores. A continuous record of climatic change can be derived from cores which include datable material and which also give evidence of climatic variations. This may be tiny microfossils, such as radiolaria, or oxygen isotope data, both of which have provided supporting

Figure 3 (a) The Milankovich astronomical cycles.
(b) The major cycles associated with the Milankovich astronomical variations.
(c) Variations in glacial ice volume derived from deep sea core isotope analysis (after J. and K.P. Imbrie, 1979)

evidence for the astronomical periodicities. It has been established that in the last 400 000 years there have been four major glacials and interglacials, including the present interglacial. The interglacials are relatively short periods of high temperature and minimum ice volumes. The glacial periods are longer, but include several swings of lesser dimensions, giving the interstadials within the glacials.

Table 3 gives a correlation table relating the glacial and interglacial sequences in Britain, Europe and North America. Both the Scandinavian and Laurentide ice sheets have repeatedly grown and disappeared, as indicated in the table. Only about 18 000 years ago, in the last major advance in the Wisconsin, the ice reached as far south as New York, while much of northern Britain was also ice covered at this time. J.T. Andrews considers that ice did not retreat north of the St. Lawrence valley between 18 000 and 80 000 years ago. The retreat since 18 000 years ago is better known, owing to a great deal of information concerning sea level changes, which are intimately linked with variations in volume and extent of ice cover. Land uplift can also indicate the main centres of the ice sheet at its maximum, as shown in figure 4. The greater the weight of ice lying on the land, the more the land will be uplifted when the weight is removed as the ice melts. The uplift data suggest that the three separate centres of the ice sheet in North America coalesced to form one major ice sheet centred over Hudson Bay; later, the centre moved south to be situated over James Bay. Ice probably advanced rather later than elsewhere in the high latitude Arctic area of north Canada, between 10 000 and 13 000 years ago, when higher temperatures allowed heavier snow falls.

In Europe the last major glaciation, the Weichselian, was nearly as extensive as the early Saalian. The Saalian ice wasted away entirely in the Eemian interglacial, but ice grew again in the Weichselian, which lasted in several phases from about 70 000 to 10 000 years ago, the last major advance being about 20 000 years ago. One of

British Isles	Northern Europe	Alps	North America	
Devensian glacial	Weichselian glacial	Würm glacial	Wisconsin glacial	latest
Ipswichian interglacial	Eemian interglacial	—	Sangamon interglacial	↑
Wolstonian glacial	Saalian glacial	Riss glacial	Illinoian glacial	
Hoxnian interglacial	Holstein interglacial		Yarmouth interglacial	
Anglian glacial	Elster glacial	Mindel glacial	Kansan glacial	
—	—	—	Aftonian interglacial	
—	—	Gunz glacial	Nebraskan glacial	↓
—	—	Donau glacial	—	earliest

Table 3 Correlation of glacial periods

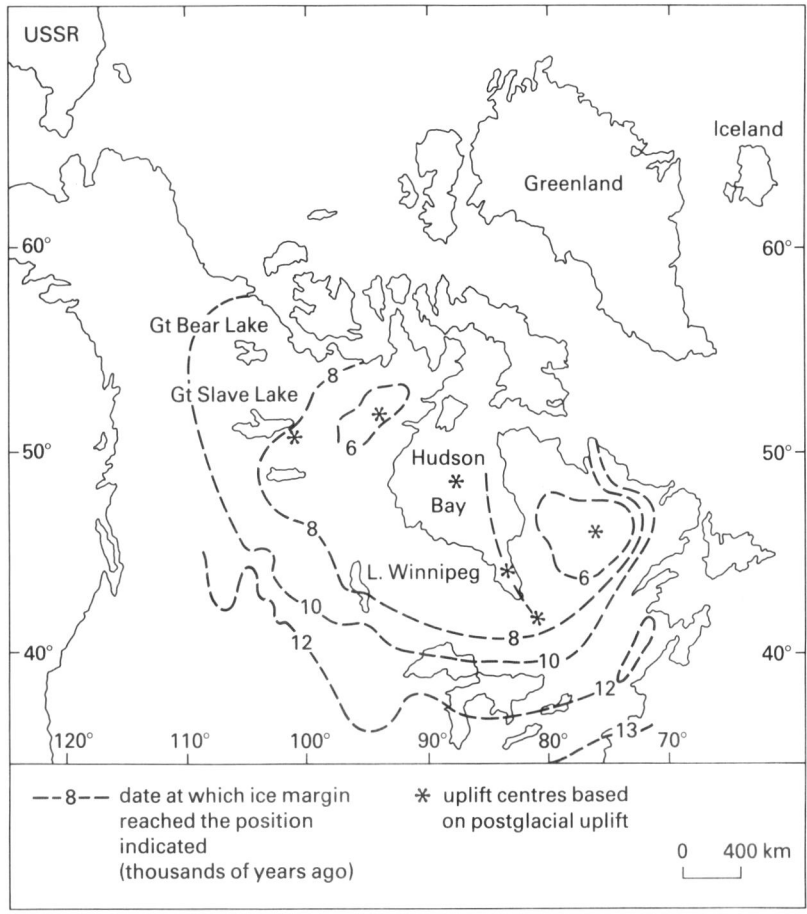

Figure 4 Land uplift centres in North America, which indicate main ice distribution and thicknesses, and times of deglaciation (after J.T. Andrews, 1975)

the characteristic features of the large northern hemisphere ice sheets was the very rapid disappearance of the ice. Much of the evidence for this comes from the last phase of retreat, which occurred rather sooner in Europe than in North America.

There is more evidence for the retreat of the Laurentide ice sheet than for its advance. Retreat from the maximum occurred mainly after 14 000 years ago. There was a period of retreat known as the Two Creeks, which ended about 11 800 years ago, and which was followed by the Valders re-advance about 11 600 years ago. Ice retreated from the St. Lawrence valley after this, with minor oscillations. At the time of the Cochrane re-advance between 8400 and 8000 years ago, the ice sheet had a radius of about 1000 km.

An important event occurred between 8200 and 7600 years ago when the Hudson Bay area was invaded by the sea; this split the ice sheet, causing it to float and break up rapidly as icebergs. In this way the Keewatin and Labrador centres were re-established. At about 6700 years ago the ice centred over Foxe Basin was split into separate masses over Baffin Island and on the Melville Peninsula. These remaining ice masses broke up rapidly until only the small Barnes ice cap remained on the plateau of Baffin Island. It now covers about 6000 km² and represents the last remnant of the great Laurentide ice sheet.

The suggestion that flotation played an important part in the very rapid break up of the ice sheets can be considered further in relation to sea level changes. Once sea level started to rise as the climate improved and ice started to melt, those ice sheets that were on low ground would be liable to be undercut by the rising sea and so float away relatively quickly. For example, when the sea invaded Hudson Bay, massive 'calving' (the production of icebergs) of the floating Laurentide ice sheet caused retreat rates of up to 600 m/year.

The Scandinavian ice sheet may have been the first to be affected in this way, as water entered the Baltic and allowed very rapid retreat as the ice floated. There is evidence in the flooding of the Baltic area by the Littorina Sea. This marine incursion occurred as the Scandinavian ice sheet was retreating and accelerated its retreat. The melting of the Scandinavian ice sheet raised sea level further so that in its turn the Laurentide ice sheet was affected by the incursion of Hudson Bay by the sea a few thousand years later than the invasion of the Baltic. This is one factor that could explain the asymmetrical pattern of glaciation, with a slow build-up and a very rapid retreat.

The build-up of the ice sheet was helped by the processes referred to as positive feedback, whereby the change in one variable leads to a change in another that enhances the change in the first. Thus when snow started to accumulate on the north Canadian uplands it reduced the area of bare ground and increased the albedo, thus causing a further lowering of temperature and less melting. Such processes can continue until a lowering of the temperature reduces the precipitation, and then their reversal is initiated. For this reason it has been argued that the zone of maximum snowfall gradually moved south as the Laurentide ice sheet grew, because it received its moisture along the southern margin. This would be another factor that would ensure its great extension into relatively low latitudes.

Ice and its behaviour

Snow to ice

Glacier ice originates as snow, which has gradually been modified by various processes, first to form 'firn' and then ice. Newly fallen snow has a density of 50 to 70 kg m^{-3} and is in the form of hexagonal crystals. The corners melt and the crystals become granular and form firn, which is literally snow that has survived one melting season. It has a density of 400 to 830 kg m^{-3}. At the latter density the air passages between the grains become sealed off and the material becomes ice, the density of which can increase to 910 kg m^{-3}. Packing of the grains is most important when there is no meltwater, but if there is, the process of ice formation is accelerated.

The time taken for ice to form from firn varies with the temperature and precipitation, becoming longer as it becomes cooler and drier. In polar areas, for example at Vostok in Antarctica at 78.5 degrees south, with a snowfall equivalent to 22 kg m^{-2} per year, and a temperature at a depth of 10 m of $-57°$ C, the age of transition to ice is about 4000 years and the depth by this time will be 100 m where the ice forms. At the other climatic extreme at Dye in the Arctic at 65.2 degrees north, with a precipitation of 490 kg m^{-2} per year and a temperature at a depth of 10 m of $-19°$ C, the depth of transition to ice was 65 to 70 m and the time taken was 100 years. In temperate areas the time and depths are much less. At Vallée Blanche in the Alps, for instance, the depth is about 32 m and the age 13 years of the newly formed ice.

One process by which snow is changed is by the formation of depth hoar. This consists of cup-shaped crystals formed by sublimation, which is the direct evaporation of water vapour from the snow crystal and its recrystallisation. Depth hoar gives rise to a layer of porous, low density snow, that is often very unstable. It can form a weak surface along which avalanches frequently break away. Depth hoar can only form in unconsolidated snow, but it can be very dangerous in mountain areas.

Glacier regimes

A glacier regime, or *mass balance*, is dependent on the relationship between accumulation and ablation. *Ablation* is a term which covers all the processes of ice loss, of which melting is the most important normally, over the whole glacier system. The values for the accumulation and ablation are usually expressed in *water equivalent* terms, and depend on the snow and ice density. The volumes of ice and snow are converted into the equivalent amount of water and expressed as a depth per unit area, which is similar to precipitation measurements. The glacier balance is calculated over one balance year, which runs from the time of least ice volume at the end of the summer ablation season, as shown in figure 5 (which applies to a single point). The method can be extended to cover the whole glacier system.

The balance varies over the glacier area and is positive in the accumulation zone above the *equilibrium line*. This line is where accumulation balances ablation. In the ablation zone there is a net loss of volume. Glacier flow maintains the volumes of the two zones by allowing ice to move down from the accumulation zone to the ablation zone. In temperate glaciers the equilibrium line coincides with the limit of the previous year's snow at the end of the summer season, which is the firn line. Figure 6 illustrates the mass balance of a glacier. It shows the area of the glacier at any given height range, and the summer and winter balances for each elevation. The difference between the two curves gives a net balance, and the total balance can be calculated by multiplying the net balance for each elevation by the surface area of the glacier at that height.

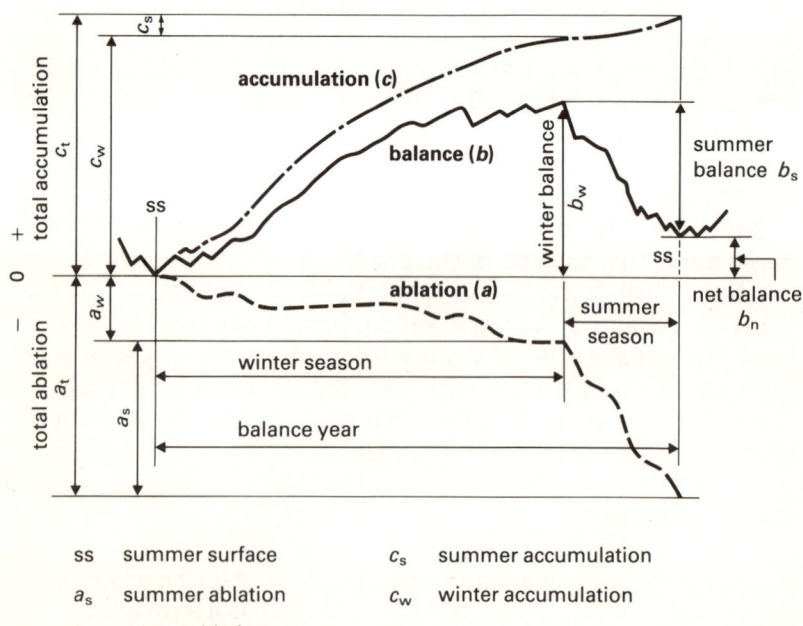

ss	summer surface	c_s	summer accumulation
a_s	summer ablation	c_w	winter accumulation
a_w	winter ablation		

Figure 5 Diagram to illustrate the elements of the glacier mass balance over the balance year (after J.T. Andrews, 1975)

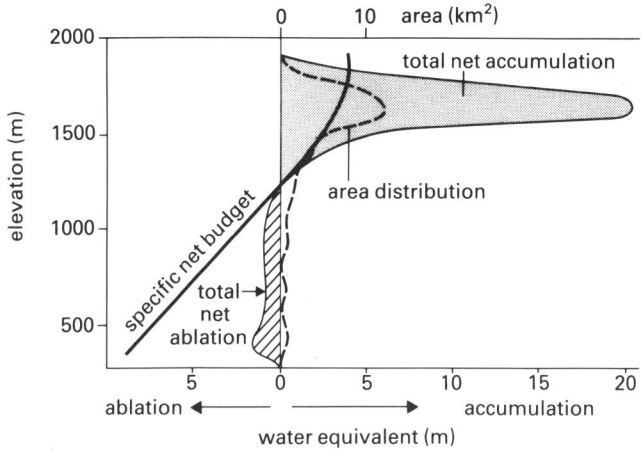

Figure 6 Diagram to illustrate the variation in ablation and accumulation in relation to glacier area and elevation, for Nigardsbreen, Norway, for the balance year 1961–62 (after G. Østrem, 1961–2)

It is rare for the accumulation and ablation to balance exactly; usually the balance is either negative or positive, and if the trend continues for several years the glacier snout will respond to the change by advancing or retreating. The variations in accumulation and ablation may be transmitted down the glacier as pulses called kinematic waves. These are bulges of excess mass that pass down the glacier more quickly than the ice itself. If there is a marked and sudden increase in accumulation at the head of the glacier, the upper part will respond in a stable way, and gradually increase in thickness. In the ablation zone, where the ice is being compressed, the response will be unstable. This leads to an exponential, or accelerating, increase of ice volume. The zone of volume increase, which is the crest of the kinematic wave, moves down the glacier at a speed about four times that of the ice velocity. Lateral extension, which is called diffusion, spreads out the bulge of ice. The final response to the change in balance, however, may take decades or centuries to reach the snout of the ice mass. The response time is the period needed for a glacier to adjust to a change in balance. Typical values of response time for valley glaciers are in the range 2½ to 25 years. For large ice sheets it is much longer, about 25 000 years. Ice sheets are relatively insensitive to changes in accumulatio

An important geomorphological effect of the mass balance of a glacier is the size of the *gross balance*, which is the sum of the positive and negative elements of the equation, regardless of the sign of the *net balance*, which is their difference. The gross mass balance is high where the accumulation and ablation are large, and

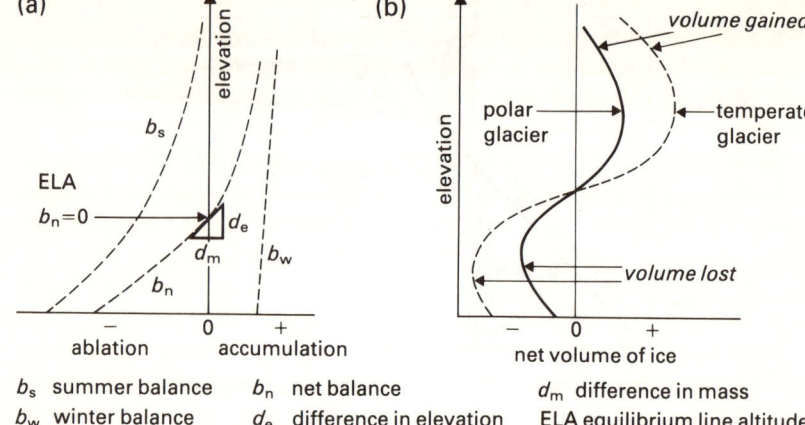

| b_s summer balance | b_n net balance | d_m difference in mass |
| b_w winter balance | d_e difference in elevation | ELA equilibrium line altitude |

Figure 7 Diagram to illustrate the glacier activity index. At the equilibrium line altitude (ELA), the slope of the b_n curve ($= d_e/d_m$) gives the ratio of mass to elevation in mm/m

low where they are small. A high gross mass balance leads to a high glacier activity index, while a low gross mass balance gives a low glacier activity index. The *glacier activity index*, as defined by J.T. Andrews, is illustrated in figure 7(a); it is the gradient of the net balance line at the elevation (height) of the equilibrium line, where the net balance, b_n, is zero and is given in millimetres of net mass balance change per metre unit of elevation (mm/m).

The values range from about 1 mm/m for high polar glaciers to about 10 mm/m for temperate glaciers. The former glaciers have low precipitation and also low temperatures, so that both accumulation and ablation are small (figure 7(b)). The glacier flow tends to be slow and the activity index, therefore, is low. On the other hand, in a temperate glacier such as those in south Iceland, the Alps and New Zealand, the precipitation is very high and the temperatures are also relatively high, so that ablation is rapid. The glacier must, as a result, flow fast to transfer the high accumulation in the upper part to the snout of the glacier across the equilibrium line. Accumulation on temperate glaciers can reach 3.5 to 4.0 m water equivalent, and summer temperatures can be +5° C, while high polar glaciers have a snowfall of less than 0.2 m water equivalent and summer temperatures of about −2° C.

Types of ice masses

Ice masses can be classified in a number of ways according to different criteria. One classification has already been mentioned in

16

the previous section. This is the classification in terms of the activity of the glacier, which depends on the gross mass balance. Another important classification is the thermal one, because the temperature of an ice mass plays an important part in its geomorphological activity.

Ice masses can be classified as temperate where the ice is at *pressure melting point*. This means that meltwater can exist at the base of the ice even if the temperature is below 0° C, as the pressure of the overlying ice lowers the melting point temperature. The situation is, however, complicated by the fact that a single ice mass may be temperate in parts and cold elsewhere. In cold ice the temperature is below the pressure melting point temperature, and the ice is frozen to the bed. In most temperate glaciers the temperature is close to 0° C throughout the ice, although normally the winter cold slowly penetrates downwards through the ice, causing a gradual lowering of temperature of about (6×10^{-4}) °C every metre downwards through the ice. The important point about temperate glaciers is that there is meltwater between the ice and its bed. The glaciers of the Alps and southern Norway are temperate in type, and meltwater issues from beneath the ice. Most larger glaciers and ice sheets are mainly cold, but they often have temperate parts, especially in deep basins and fairly near, but not at, their margins.

Cold glaciers remain frozen to their beds and meltwater streams flow on the surface. These ice masses make much more effective barriers for ice-dammed lakes, as water cannot penetrate easily through them. Examples of ice-dammed lakes occur around the edge of the Barnes ice cap on Baffin Island. Large ice sheets, such as the Greenland and Antarctic ice sheets, have complex temperature patterns, which depend on the climate, the ice thickness and geothermal heat supplied from the earth beneath the ice. Basal shearing, whereby the glacier moves over its bed and friction is generated in the lowest layers of ice, also produces heat. There is, as a result of these effects, a rise of temperature with depth in large ice sheets. For example, at Vostok in Antarctica, the temperature can be as low as −60° C on the surface near the centre of the ice sheet where temperatures are very low and precipitation small. At a depth of about 4600 m below the surface, where the ice is estimated to be about 200 000 years old, the temperature is above −20° C. Near Byrd station, where the ice is over 4000 m thick and 30 000 years old, the temperature is above the pressure melting point, so that melting can occur in the deep basins beneath the ice sheet as shown in figure 8. Temperatures can also reach pressure melting point near the margins of some ice sheets. The temperature of the basal ice is very significant as far as its movement is concerned, and this aspect

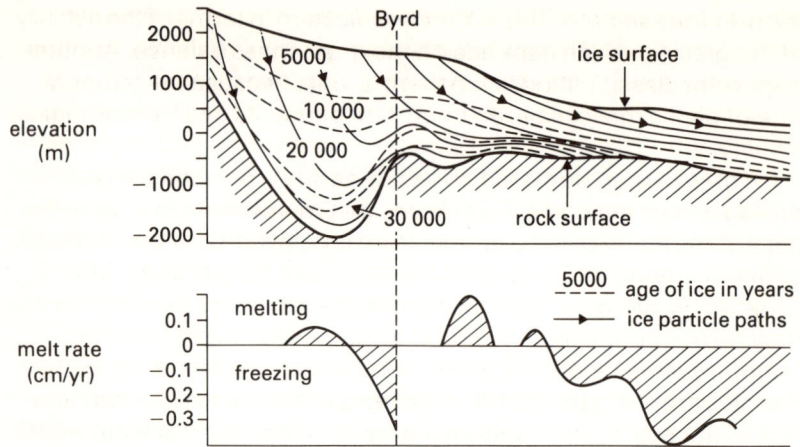

Figure 8 Profile across the Antarctic ice sheet through Byrd station to show the ice thickness, ice particle paths, the age of the ice and the pattern of melting and freezing at the base of the ice sheet (after W.F. Budd, 1970)

will be considered further in connection with glacier movement.

The classification of ice masses in terms of their size and position is also relevant to the geomorphologist. At one end of the scale are the smallest ice masses, which have been called either niche, wall-sided or cliff glaciers, and at the other end are the continental ice sheets. There is a continuum of features between snow patches, various forms of niche glaciers and well developed cirque glaciers. In order to be classified as a glacier the ice must move, and it must be true ice as defined at the beginning of the chapter, having no interconnecting air passages. Thus snow patches that occupy nivation hollows cannot be called true glaciers, even though the snow may last several years and does creep very slowly downslope. The full sequence includes the following:

1. niche, wall-sided or cliff glacier
2. cirque or corrie glacier
3. valley glacier: (a) Alpine type
                    (b) outlet type
4 .diffluent glacier
5. transection glacier
6. regenerated glacier
7. piedmont glacier
8. floating ice tongue or ice shelf
9. mountain ice cap
10. glacier cap or ice cap
11. continental ice sheet.

Niche glaciers occupy gullies in the hill side or lie on shelves on which snow can accumulate until it is thick enough to become ice and flow. Typical cirques are armchair-shaped depressions in the hills and *cirque glaciers* are rounded, steep ice masses. Skauthoebreen in the Jotenheim of Norway, for example, is about 1 km long and has a slope of about 26 degrees. The glacier base has a circular long profile and occupies a hollow. The backwall is steep, and a bergshrund, which is a large crevasse close to the backwall, separates the slowly moving ice from that adhering to the backwall. Figure 9 illustrates the characteristics of this typical cirque glacier.

Valley glaciers are divided into various types. Some form when the snow line is lowered enough for the ice to move out of the cirques into the valleys below. These are referred to as the Alpine type, as there are many glaciers like this in the Alps, such as the Aletsch glacier. Several cirques may feed one valley glacier, which is surrounded at its head by nunataks, which are rock outcrops surrounded by ice or snow, in the form of Alpine peaks, of which the Matterhorn is a good example. Another type of valley glacier is one which drains from a mountain ice cap or ice sheet. These glaciers are called outlet glaciers. Many glaciers of this type flow from the ice caps of Iceland, Norway and the ice sheets of northern North

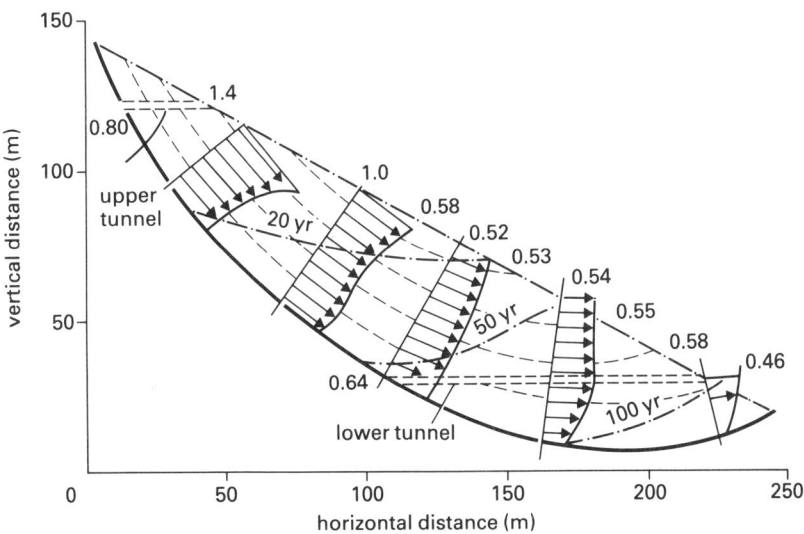

Figure 9 Profile through Skauthoebreen to show the nature of the internal flow and deformation of the ice layers through time. The ice velocity is indicated by figures in cm per day. The parallel arrows indicate direction and speed of flow through the depth of the ice at the positions indicated by the straight lines. Broken lines indicate flow paths through the ice mass. The dot–dash lines indicate age of the ice in years. The tunnels were dug to reveal internal structure and movement (after J.G. McCall, 1960)

Thorsbre, on the extreme left, is a very steep glacier flowing from Jostedalsbreen to join Austerdalsbreen, an outlet glacier, in the foreground. The very high, steep gneiss walls of the valley indicate the great erosive capacity of fast-flowing, thick and active glaciers. The medial moraine can be seen and ogives, which are annual bands, indicate the flow each year.

America and Greenland. There are also many long valley glaciers in the high mountains of Asia.

Some of the types of glaciers listed can also be included within the general class of valley glaciers, including regenerated glaciers, diffluent and transection glaciers. *Regenerated glaciers* form from ice that has avalanched from higher areas into a valley where it is welded again into glacier ice. The Ventisquero Negro of the Argentinian Andes is an example. It is about 5 km long and gets its name, which means black, from the large amount of surface dirt on it due to the avalanche activity which forms the glacier. *Diffluent* and *transection glaciers* form where the snow line falls so far that the main valley can no longer evacuate all the ice fed to it from the snowfields. A diffluent glacier flows out from the main glacier as a distributary, while a transection system forms where the ice escapes over many cols. Examples are found in Spitzbergen and Greenland, and they once occurred in Scotland, on Rannoch Moor, for instance. The blockage of the outlet of a valley glacier by ice from other sources may also cause diffluence to occur.

Piedmont glaciers and *floating ice tongues* form where the snow-line is low enough for the ice to reach the lowlands. Piedmont glaciers are land-based, while glacier tongues occur where the ice advances into the sea, a fjord or a large lake. The Malaspina glacier in Alaska is a well-known piedmont glacier. Its broad snout occupies a hollow that lies well below sea level, so that the basal ice must move upwards towards the glacier margin. The actual margin of the glacier is stagnant ice, on which large trees flourish on the surface debris. Many floating ice tongues occur around the Greenland ice sheet, where they occupy narrow, steep fjord valleys. The ice will float when the water depth is sufficient. The Steensby glacier in north Greenland, for example, floats for about 8 km of its length of 48 km. Its floating section is estimated to be only 75 to 100 m thick, as its frontal ice cliffs rise only 15 to 18 m above sea level. The break-up of floating glaciers gives rise to icebergs in the sea.

Ice shelves are more characteristic of the Antarctic, where the large Ross and Weddell ice shelves cover large areas. The Ross ice shelf covers 550 000 km² and is 300 to 400 m thick. It is composed mainly of firn. The ice spreads out and moves under its own weight at rates between 4 and 844 m/year. The ice 'calves' to form the very large tabular icebergs, up to several kilometres in length, that are found in Antarctic waters. Over most of its area the shelf melts from below.

Mountain ice caps occur on upland plateaux, and the Jostedalsbreen in Norway and Vatnajökull in Iceland provide good examples. These ice caps give rise to a number of outlet glaciers. They are in relatively low latitudes and are probably temperate throughout. The Barnes ice cap in central Baffin Island is an example of an *ice cap* or *glacier cap*. Such ice masses form on fairly low ground, but only in high latitudes in the Arctic. The Barnes ice cap has no vigorous outlet glaciers and it is probably frozen to its base over much of its area, which is 5900 km². It is a sluggish ice mass and moves very slowly at only a few metres per year, with a low activity index.

At present there are only two large *ice sheets*, the Greenland and Antarctic ice sheets. The Antarctic ice sheet is by far the largest ice mass on the earth now, and it has probably existed for tens of millions of years as already discussed. Its disappearance cannot be foreseen in the future, as it is a stable and massive ice sheet occupying an area around the south pole where temperatures are very low.

Ice flow

It has long been known that ice can flow. Observations on glaciers show that the ice flows fastest at the centre because the friction of the rock walls slows down the ice flowing along the glacier margins. Where the ice is thin and rests on a steep slope, however, the glacier may flow as a solid body; this type of flow is referred to as *block-schollen* flow. It is most likely to occur in ice falls and just below them. Measurements on Austerdalsbreen in Norway showed that there was a very rapid reduction of speed below the ice fall where the velocity was between 1000 and 2000 m/year, while in the gently sloping glacier snout area the flow was reduced to between 50 and 100 m/year, with the velocity along the margin of the ice about a fifth of the centre speed. The down-valley velocity of glacier flow also decreases downwards through the ice, although it can be a considerable proportion of the surface velocity even on the bottom.

In general ice will have a downward component of movement in the accumulation zone. It will reach its most rapid velocity at the equilibrium line, moving parallel to the surface. The flow is slower near the snout in the ablation zone, where there is an upward component of flow. These findings can be related to the changes of mass of the glacier through accumulation and ablation as shown in figure 10.

The study of glacier flow has been largely carried out on the theoretical side by physicists, who have compared ice to viscous and plastic materials. In such deformable substances the strain rate is the change in length of a measured distance over a period of time. It is related to stress, which is the force exerted per unit area. A considerable advance in the theory of ice flow was made by suggesting that ice behaves as a perfect plastic substance, with a yield stress of about 100 kPa, or 1 bar. The yield stress is the stress necessary to initiate deformation. The strain rate is the response of the ice to the stress applied to it. The flow law of ice relates these two variables as shown in figure 11. The strain rate, ε, is given as $A\tau^n$, where τ is the shear stress and A and n are constants. The shear stress is the force exerted laterally on the ice. The value of A depends on the temperature. If the temperature is $0°$ C the value of A is 0.165, for $-5°$ C A is 0.054, and at $-10°$ C it is only 0.017, falling as low as 0.0015 at $-20°$ C. Thus it varies over two orders of magnitude. The values of n vary between 1.7 and 4.2, and depend largely on the

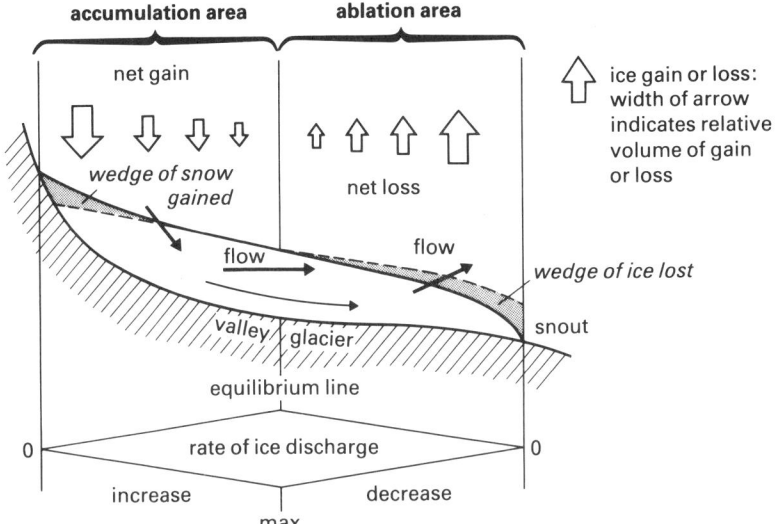

Figure 10 Diagram to illustrate the relationship between accumulation, ablation, ice discharge and ice flow in a simplified valley glacier (after B.S. John)

orientation of the ice crystals, being higher as the crystals become better orientated.

The shear stress, which acts parallel to the bottom of the glacier, is given by the relationship $\tau = \rho g h \sin \alpha$, where ρ is the ice density in kg m^{-3}, g is the acceleration of gravity in cm s^{-2}, h is the ice thickness

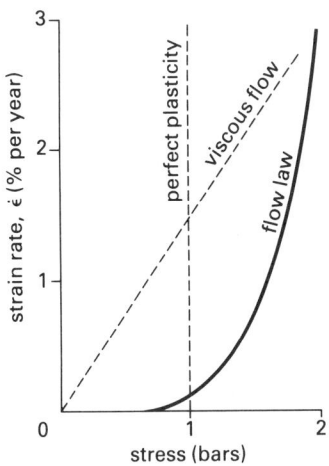

Figure 11 Diagram to illustrate the relationship between stress and strain for viscous flow, perfect plasticity and the flow law of ice

in metres and α is the ice surface slope angle measured from the horizontal. As shown in figure 12, it is the difference between the force acting vertically down, $\vec{CB}$, and the force acting normal to the glacier surface, $\vec{CA}$, and is thus related to $\vec{AB}$. This relationship has several important consequences. Firstly, the value of the shear stress can be established from measurements of the ice thickness, h, and the surface slope, α. The value of shear stress usually lies between 50 and 150 kPa (0.5 to 1.5 bars). These values are close to the yield stress of ice, the stress at which it starts to deform, according to the theory of perfect plasticity. If perfect plasticity is assumed then $h = \tau/(\rho g \sin \alpha)$. This implies that a measurement of glacier thickness can be approximately associated with surface slope. A thin glacier will have a steep surface and vice versa. The relationship allows the reconstruction of former glacier profiles, and these can then provide valuable material for the calculation of former ice volumes and loads, and can also be related to lateral moraines.

If a glacier deforms by simple shear the flow is called laminar. In this type of flow the flow lines are parallel to the surface, and the maximum change of flow rate occurs near the base. Allowance must also be made for the effect of the side walls of a normal valley glacier.

In most parts of the glacier the situation is more complex, and a longitudinal extension or compression is superimposed on the shear. J.F. Nye has analysed this situation. If the glacier is assumed to be very wide so the side wall effect can be ignored, the variation in velocity down the glacier and variations in accumulation and ablation are taken into consideration in deriving the theoretical flow patterns. Two types of flow have been recognised, extending and compressing flow. Extending flow will occur in the accumulation zone, where the volume of ice discharge is increasing down-glacier, and where the glacier bed is convex. Compressing flow will occur in

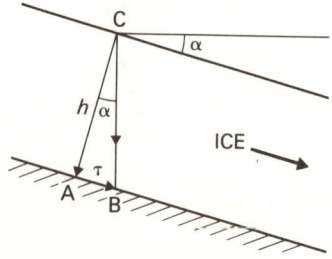

Figure 12 Diagram to explain shear stress in a glacier in relation to the ice thickness and glacier slope

the ablation zone, where ice discharge is decreasing down-glacier, and where the bed of the glacier is concave. Extending flow will occur at the top of ice falls, while compressing flow will normally occur at their base and near the snout of the glacier.

The significance of these flow types is related to the development of planes of weakness within the ice which are associated with them. Figure 13 indicates the pattern of flow. The important planes of weakness are those that slope up tangentially from the glacier bed in a down-glacier direction in the zones of compressing flow, because these lead to enhanced erosion by the glacier. In the zones of extending flow, the planes of weakness slope down tangentially from up-glacier and erosion is less likely to occur in these zones.

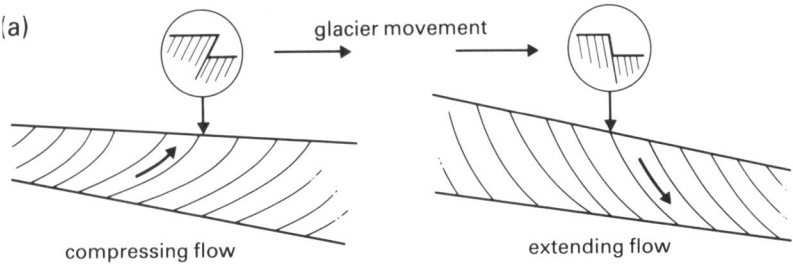

(a) glacier movement

compressing flow   extending flow

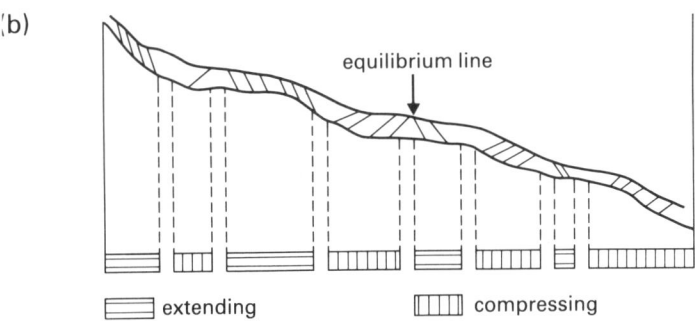

(b) equilibrium line

☰ extending   ⊞ compressing

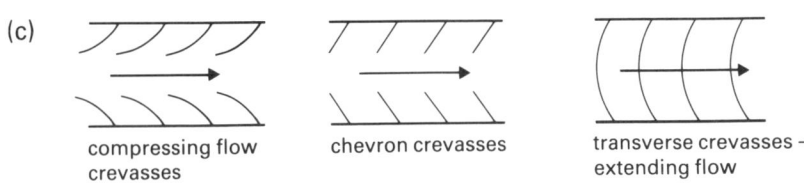

(c)

compressing flow   chevron crevasses   transverse crevasses –
crevasses   extending flow

Figure 13 Diagram to illustrate extending and compressing flow and the associated crevasse patterns (after J.F. Nye, 1952)

25

This point will be considered further in relation to glacier erosion (p. 36). The types of flow can often be recognised by the crevasse pattern (figure 13(c)). Transverse crevasses, which are convex up-glacier, are found in zones of extending flow, while in the zones of compressing flow the crevasses tend to splay out from the centre, with a convex down-glacier pattern. These patterns can often be recognised on aerial photographs.

Most glaciers flow relatively steadily, although many flow rather faster during warmer and wet weather, but a few glaciers and ice sheets are unstable. These are referred to as *surging glaciers*, because of their periodic and usually short-lived but very rapid advances. Surges can affect ice masses of different sizes, from small glaciers to ice caps. A surge is not the same as a kinematic wave, because in a surge the whole mass of the glacier moves, while a kinematic wave is an accelerated movement of a bulge of ice down the glacier. Only a small number of glaciers surge. However glaciers that do surge do so fairly regularly. The cycle consists of a short-lived rapid advance of the snout, followed by a period of stagnation and decay. This, in turn is followed by a slow build-up of the upper part of the glacier surface, until instability causes another sudden surge.

A surging glacier can move very rapidly, often at a hundred times the normal flow, or up to 100 metres per day advance at the snout. The snout may advance several kilometres down-valley over a period of a few months during a surge. Ice is transferred from the reservoir area at the head of the glacier suddenly to the snout. The periodicities are often 50 to 100 years between surges. As the surge builds up, the upper part of the glacier thickens due to high values of upward flow in the accumulation zone, while the lower part is thinned by ablation of the virtually stagnant ice of the previous surge. The boundary between active and stagnant ice advances down-glacier during quiescent times. Once the profile reaches a critical state, the glacier ice can no longer remain relatively static and a sudden forward movement is initiated, leading to the surge of the glacier, and rapid sliding on the bed probably occurs on a build-up of basal water. There is no evidence of surges in any glacier that is known to be permanently frozen to its bed.

Observations are now being carried out on the Medvezhiy Glacier in the Pamirs of Soviet Central Asia, which will provide information essential to the full understanding of the origin and mechanism of surging glaciers. This glacier has an area of 25 km² and its tongue is about 8 km long. The diagrams in figure 14 illustrate the velocity pattern and the changes in the profile during the surge cycle. Glacier and ice sheet surges are now thought to explain some hitherto

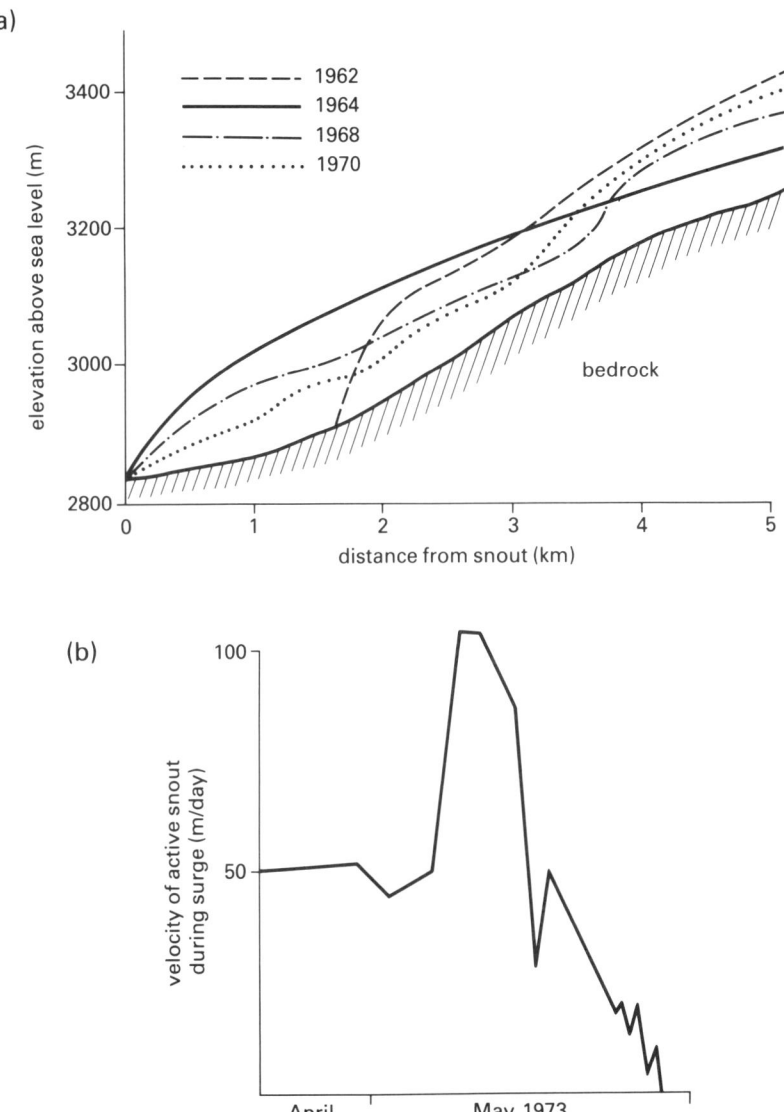

Figure 14 (a) Profiles to illustrate the characteristics of surging glaciers at various stages of the surge cycle for the Medvezhiy Glacier. (b) The velocity of the glacier snout during a surge in April and May, 1973 (after W.S.B. Paterson, 1981)

unexplained features of glaciation during the last ice advance, such as the great distance ice travelled down the east coast of England, reaching the north coast of Norfolk.

Glaciation in the uplands

Ice in the uplands is predominantly an erosive agent, creating the familiar features of the uplands of the Lake District and Snowdonia, for instance. These features include the cirques or corries, arêtes, peaks and deep, U-shaped valleys with their long lakes and valley steps or riegels. The nature of ice flow provides information concerning the processes of erosion, which also depend on the type of ice and the character of the land surface over which it is flowing. Processes of glacial erosion will be considered first and then the features to which it gives rise will be examined.

Glacial erosion

The ability of a glacier to erode is intimately associated with conditions at its bed and sides. This in turn depends on the temperature regime of the glacier. Cold glaciers that are frozen to their beds, thus preventing bed slip, are less efficient than temperate glaciers as erosional agents.

The type of flow influences the nature of glacial erosion. The slip planes associated with extending flow tend to inhibit erosion. Material in the ice is carried down to the glacier bed, thus protecting the ground from erosion. In areas of compressing flow, on the other hand, the slip planes carry material up into the ice, exposing the bed to further erosion (see figure 13(a)). This is a positive feedback process, whereby developing hollows gradually become deeper as the concavity increases in curvature. Nye and Martin, however, have suggested that the process cannot go on for ever. A limit to the erosion will be reached when the slip planes fit the glacier bed. There is no limit to the convex curvature, although when it becomes very sharp it will cause the glacier to become so thin that plucking could occur on the rock steps formed by this means.

There are two processes involved in the slip of a temperate glacier over its bed. One is *regelation flow*, which is most effective when the ice encounters small obstacles. The increase of pressure on the upstream side of the obstacle causes pressure melting, and the meltwater flows to the downstream end of the obstacle where it refreezes. The other process is *enhanced plastic flow*. A glacier, on meeting a large obstacle of more than 1 m in size, flows faster due to the increase of pressure that causes increased plastic deformation. The sliding velocity of the glacier is determined partly by the size of the obstacles on its bed. Weertman has proposed a complex equation to calculate the controlling obstacle size which depends mainly on the average shear stess, the values of A and n in the flow law equation, and the thermal conductivity of the bedrock. Obstacles of the critical controlling size will create the maximum hindrance to glacier sliding.

It has been shown that water-filled cavities at the base of the ice are important in the sliding process. A gravel layer between the ice and its bed will reduce the sliding velocity. Observations have shown that in many glaciers the basal sliding rate is a large proportion of the total rate of movement, reaching 90 per cent in some instances, as in Skauthoebreen which is a small cirque glacier in the Jotunheim, Norway.

There are two main processes of glacial erosion. These are *abrasion* and *plucking*. Abrasion is brought about by the dragging of particles held in the base of the glacier over its bed. The amount of abrasion is proportional to the size of the particles embedded in the ice. Abrasion is effective in smoothing and striating the bedrock over which the ice is moving. One indication of its effectiveness is the milky appearance of glacial meltwater due to its high content of rock 'flour'. Observations have confirmed the effectivenesss of glacial abrasion. For example, experimenters placed blocks of different kinds of rock in the path of the Breiðamerkurjökull glacier in Iceland. After 9.5 m of basal ice had passed over these test blocks, they were found to be striated. Marble blocks had been lowered 3 mm by studs placed in the basal ice and basalt ones had been lowered 1 mm.

Abrasion will also be most effective where the particles are moving towards the glacier bed. The significance of the ice velocity in abrasion erosional processes is that faster flowing ice carries more particles across each section of the glacier bed than slower moving ice does.

Ice thickness is important, as the greater the ice thickness the higher the pressure exerted on the glacier bed. At a critical point, however, the friction between the particle and the bed becomes high enough to retard the movement of the particle, which will eventually lodge on the bed. The presence of a layer of water on the glacier bed can reduce abrasion, by reducing the effective pressure on the bottom of the glacier. The relative hardness of the particles and the bedrock also influences the effects of abrasion. Hard particles and a soft bed produce optimum conditions for abrasion. It is also influenced by the processes that produce the particles within the glacier on which it depends. Thus frost action, by breaking up rocks above the ice, helps to provide the larger tools that are the more effective agents of abrasion. When the particles are too fine abrasion is reduced, unless the small particles are washed away when they become too fine to be effective agents of erosion. Water to wash them away thus helps to make abrasion effective.

Abrasion is most likely to be effective on the up-glacier side of rock protruberances in the glacier bed, while plucking is usually more effective on the down-glacier side where cavities occur between the ice and the bedrock. This produces the characteristic form of roche moutonnées, which usually have a smoothly rounded and striated upper surface facing up-glacier, while their down-glacier side is rough and plucked, with angular rocky outcrops.

Plucking is rendered more effective where the rocks are first prepared. Preparation of the glacier bed for plucking to be effective is associated with the development of joints and other planes of weakness within the rocks. One method by which this can be achieved is by frost action due to a periglacial climate. Freeze–thaw action will be most effective in the lower parts of the valley where the moisture content is higher.

Another possible process is associated with pressure release. As glacial erosion proceeds rock is replaced with ice, which has a lower density. This process reduces the pressure on the rock of the glacier bed and allows pressure release jointing to develop. It is a well known process in quarries and probably operates beneath valley glaciers and cirques.

The process of plucking occurs when the frictional drag between the ice and the bedrock is overcome. This is most likely to occur in the lee of rock hummocks where cavities occur between the ice and its bed. The glacier ice deforms rapidly on the up-glacier side of the hummock and refreezing takes place on the down-glacier side. The loosened rocks can then be frozen into the ice and removed by plucking.

Unless the rocks are first loosened by one of the processes already mentioned, plucking will not be very effective. It is probably most effective early on during a glacial advance when there is more prepared material available. The pressure release mechanism can, however, take place under long continued erosion. It is also a positive feedback process, as the deeper the glacial erosion the greater the pressure release will be likely to be. The plucking process gives rise to the rough, angular shape of the down-glacier side of roche moutonnées on a small scale. On a larger scale it creates the steep and rough surfaces of valley steps in a glacial valley.

The processes of abrasion and plucking cannot go on continuously by positive feedback, as thresholds occur which set a limit to their operation. Figure 15, which is partly derived from the work of G.S. Boulton, illustrates the relationship between the effective normal pressure of the ice, the abrasion rate and the ice velocity. The

(a)

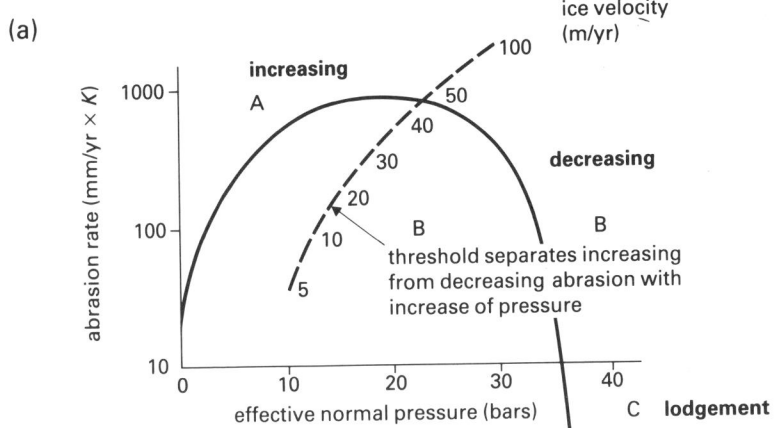

(b)

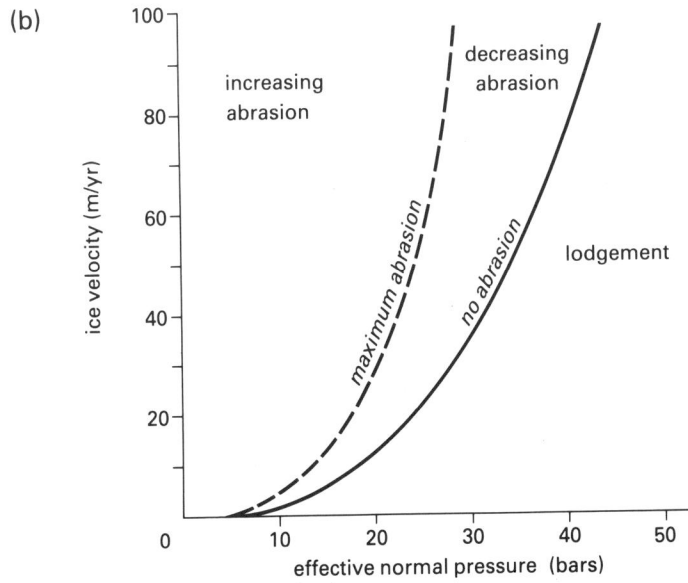

Figure 15 Diagrams to illustrate the relationships (a) between erosion rates and effective normal pressure, and (b) between ice velocity and effective normal pressure. Zones of increasing and decreasing abrasion are indicated and also the lodgement zone (partly after G.S. Boulton)

effective normal pressure is related to the ice thickness and the permeability of the glacier bed, which determines the water pressure at the base of the ice. The diagram shows that as the ice velocity increases so the abrasion rate increases with increasing

31

effective normal pressure, but it then decreases, as effective normal pressure rises above the critical threshold. Beyond this value the abrasion rate rapidly decreases until eventually lodgement takes place (i.e. the abrading particles become lodged in the rock of the glacier bed). Thus thicker, fast moving glaciers tend to be more effective agents of erosion up to a certain point. It is these glaciers that produce the features characteristic of glaciated uplands.

Cirques

A typical cirque is an over-deepened, rounded hollow in the hillside. Blea Water, beneath High Street in the Lake District, is a typical corrie basin (Figure 16). It is very similar in form to Skauthoebreen in Norway, which still contains a cirque glacier (see figure 9). This cirque glacier has been extensively studied (see figure 9). W.V. Lewis and his co-workers describe how the ice moves over the approximately circular arc of the long profile of the cirque bed to form its smooth floor by abrasion. The ice slopes at 26 degrees and has a maximum thickness of about 50 m. Measurements of the flow on the surface and in tunnels drilled through the ice (see figure 9) show that the ice is moving mainly by rotation around a circular arc, with 90 per cent of the surface flow occurring at the base. The sole of the glacier showed a 30 cm thick regelation layer (where the ice has melted and re-frozen) where the ice was thickest. The upper tunnel shown on figure 9, dug through the ice near the headwall, revealed that here plucking was occurring, with frost shattering in the bergschrund. The radius of curvature of the arc of the glacier bed was about 240 m, a figure similar to that of the long profile of the Blea Water corrie, as shown in figure 16(b).

A study of the morphometry of corries shows that their long profiles can be fitted by a curve of the form $y = k(1 - x) e^{-x}$, where y is height, x is horizontal distance and k is a constant, the value of which varies from 0.5 for open hollows with no reserve slope to 2.0 for deeply cut corries. Not all corries are associated with lakes held back by reversed slopes. The value of k could be related to the glaciological regimes in the mountains where the corries form.

The orientation of corries is also affected by the climatic regimes. In the British Isles most corries face between north-west and east, with the north-east quadrant dominant. The elevation of the corrie lip is also related to the amount of precipitation, a feature that is well exemplified in the corries of Snowdonia, where those furthest west are lowest, owing to the highest precipitation occurring in the west. They get progressively higher eastwards.

Cirque glaciers are usually not thick enough for plucking to occur

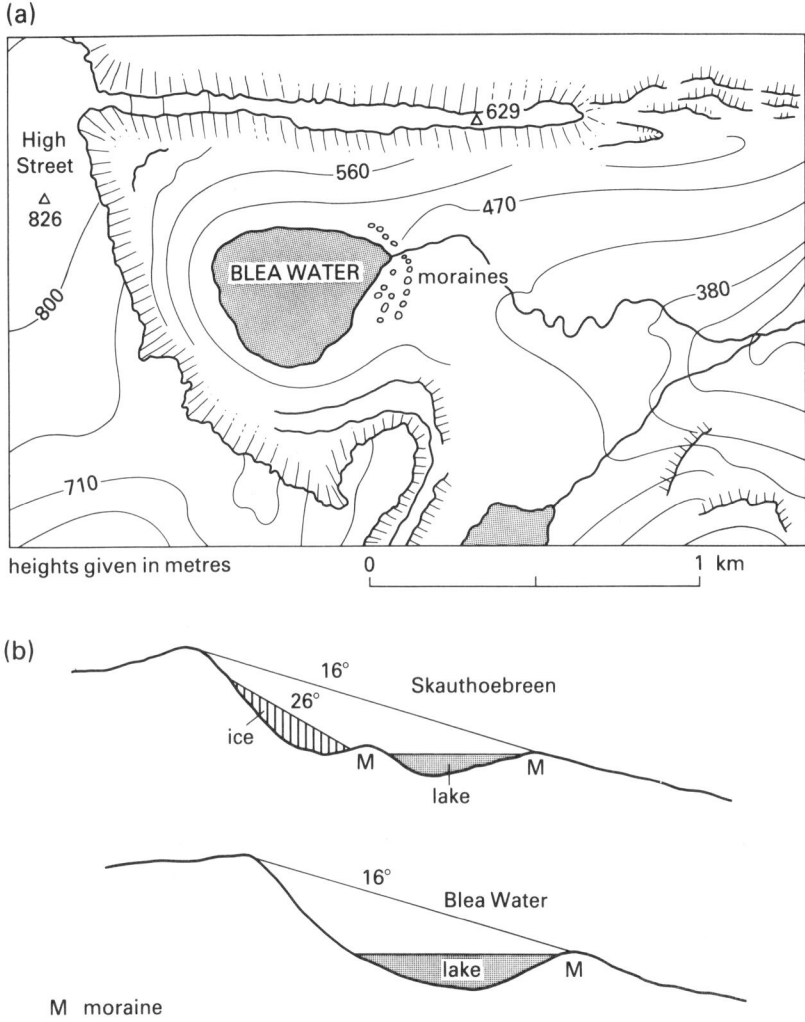

Figure 16 (a) Map of Blea Water in the English Lake District. (b) Profiles through the Blea Water and Skauthoebreen cirques

effectively beneath the ice. Nevertheless the stress is sufficient for abrasion to occur, with the help of rocks incorporated into the ice from the backwall through the bergschrund. The ice thickness and velocity reach their maxima at the equilibrium line, and here abrasion will be most effective, with the ice flowing parallel to the bed and scouring the typical hollow. Upward movement occurs near the glacier snout, where moraine usually accumulates.

Looking down into Blea Water corrie from High Street in the Lake District. The long lake of Haweswater, which has been lengthened by damming for water supply, can be seen in the distance. The zone III moraines between Blea Water and Haweswater are visible in the centre of the view.

Glaciated valleys

One of the most characteristic features of glaciated valleys is their U-shaped cross profile, but they also often have an irregular long profile. Reference to figure 15, which shows the relationship between abrasion rate, effective normal stress and ice velocity, illustrates how the U-shaped profile is likely to form. The abrasion rate increases with the effective normal pressure and the ice velocity. The thickness and speed of flow of most valley glaciers is high enough for cavities to occur beneath the ice, which increases the likelihood for the critical conditions for crushing and plucking to take place.

If the valley profile were originally V-shaped the effective normal pressure will increase linearly towards the floor of the valley. The maximum pressure at the bottom of the valley could fall into any of three critical zones, A, B or C as shown on figure 15. If it all lies within the zone of increasing abrasion, A, then the maximum erosion will occur at the bottom, producing a U-shaped form. If the maximum pressure is in the zone of decreasing erosion, B, then the maximum erosion will occur some way up the valley side, and this will tend to produce a flatter U-shaped form. In the extreme case lodgement will

34

Blea Water in the Lake District can be seen in the bottom right of the picture, and the steep back wall of the corrie leads up to High Street at the top left of the picture.

occur on the valley floor, but even then the tendency will be to widen the valley towards its floor.

A study of the form of the cross profiles of glaciated valleys often shows that they approximate to a parabola of the form $y = ax^b$, where y is the height above the valley floor and x the horizontal distance from the mid-point of the valley; the exponent b is 2 for a true parabola. The closer the calculated exponent b is to 2, the closer the curve approaches a true parabola. The values for the valley of Grisedale in the Lake District and the Nantlle valley in Snowdonia both approach 2 closely.

The long profile of a glaciated valley has a characteristically stepped form, with rocky riegels between basins, many of which are deep. The type of flow associated with extending and compressing flow is likely to enhance any irregularities in the glacier bed by positive feedback, thus deepening the hollows and making the riegels sharper. Nye and Martin have suggested that the deepening process will cease when the slip planes fit the floor of the glacier long profile. The process of over-deepening can, however, go on a long way before this form is reached.

Some of the deepest glacial troughs are those that were occupied by fast flowing ice that had a free outlet to the sea. The Sognefjord in Norway and the Inugsuin Fjord in Baffin Island are good examples, and the latter is illustrated in figure 17(a). The Sognefjord is 1308 m deep (Figure 17(b)) and its walls rise about 1800 m above sea level, giving it a total depth of more than 3000 m. The deepest known fjord is Vanderfjord in Antarctica, which is 2287 m deep. Fjords characteristically have shallow sills near their mouths. The sills are probably at least partly the result of the flotation of the ice.

A classification of glacial troughs has been suggested by D.L. Linton. He recognised four types. The first is the *Alpine* trough,

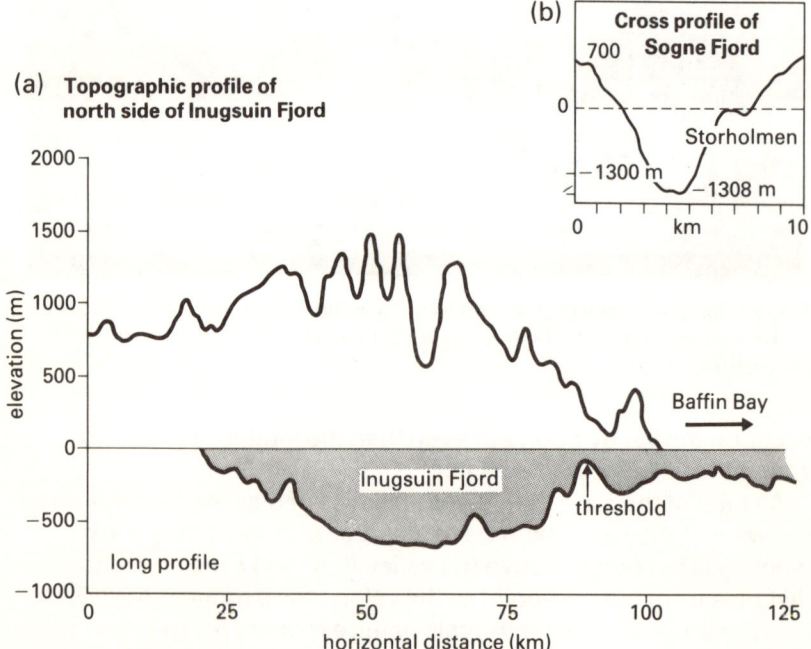

Figure 17 Long profiles of Inugsuin Fjord in Baffin Island and a cross profile of Sognefjord, Norway, at Storholmen

which usually heads in a cirque or series of cirques, and the glacier occupies the previous valley. These are common in the Lake District and Alps. The second type is the *Icelandic* type. They are formed by the outlet type of valley glacier, and are common in Iceland and Norway around the Vatnajökull and Jostedalsbreen, respectively. There are examples in Britain in the Grampian region of Scotland, where glaciers flowed out from a plateau gathering area.

The third type is the *composite* trough, of which there are five subdivisions. The first is a simple diffluent trough, of which Bishopdale in North Yorkshire and Strath Nethy in Scotland are examples. In this type the ice escapes over a col in the valley side and lowers the col and the valley into which it flows out. The second type is the multiple diffluent trough type where several side valleys are over-deepened by distributary glaciers, as around Loch Fyne. The third type is the simple transfluent trough. The Nant Ffrancon pass in Snowdonia is an example where the former watershed has been very deeply eroded and nearly eliminated. The fourth type is the multiple transfluent system, for example at the head of Glen Falloch, where ice escaped over many cols – one enabled ice to flow into the Loch Lomond valley. The fifth type is the radiative dispersal system. The radial pattern of valleys and lakes in the Lake District provides one instance. The fjordland area of southern New Zealand and southern Norway provide other examples. In all these areas ice spread out in a radiating pattern along pre-existing valleys and also cut across watersheds to give the radial pattern.

The fourth main type of trough is the *intrusive* or *inverse* trough. In this instance the ice was constrained to flow against the pre-glacial drainage. The troughs occur where powerful ice pushes up into hilly areas far from the ice source. The iceways of the Dee and Mersey Valleys on either side of the Wirral peninsula provide examples, which are now deeply buried in drift. They are illustrated in figure 18. The trough valleys of the Finger Lakes region of New York state provide other striking examples, which in this case were cut by the main Laurentide ice sheet impinging against the Allegheny Plateau.

In the production of the third and fourth types of trough the processes of glacial diffluence and transfluence play an important part. These processes can cause considerable modifications of pre-glacial drainage patterns. They account for many of the drainage peculiarities of the Scottish Highlands, for example.

Glacial deposition in the uplands is usually in the form of moraines. These hummocky areas of unsorted glacial till mark the positions at which the ice halted for some time or re-advanced a short distance during final retreat. Present-day glaciers in the hills

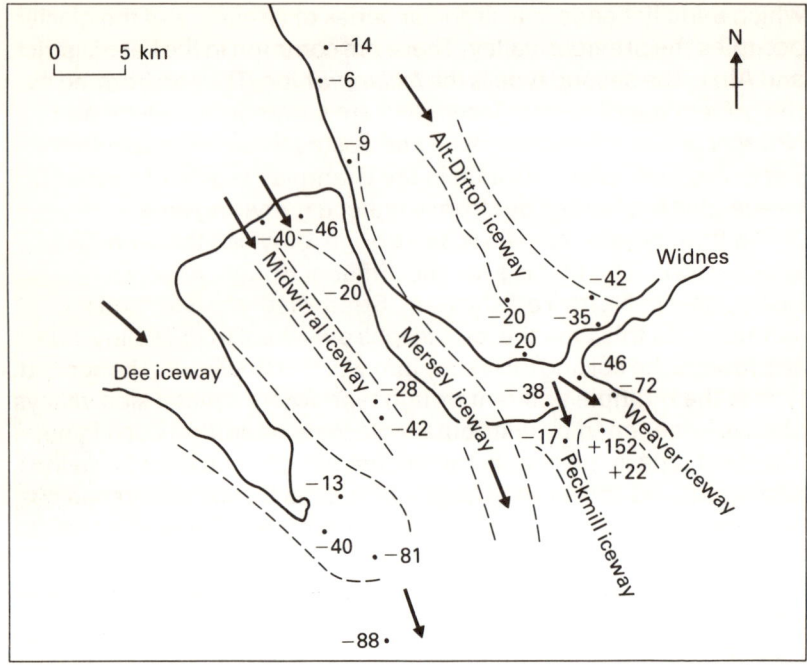

Figure 18   The intrusive troughs of the Wirral Peninsula area. The figures give the actual elevation of the bedrock in metres (− values are below sea level and + values above), to indicate depth of glacial scouring in the intrusive troughs (partly after Gresswell)

usually have conspicuous surface moraines, including lateral, medial and terminal moraines. Many of these features are, how-ever, ice-cored, and when the ice finally melts they will no longer be so conspicuous.

Even a thin covering of debris is enough to prevent the ice beneath from melting, and thus debris-covered ridges are often conspicuous on the glacier; these are called active moraines. When the ice has finally melted the debris forms inactive moraines. These can be used to date the retreat phases of the glacier. They can be dated by various methods, such as lichenometry for the more recent moraines. Other types of glacial deposits will be considered in the next chapter.

Meltwater effects

Temperate glaciers are on the whole more effective agents of erosion in the uplands than cold glaciers, as they slide effectively across their beds. Meltwater also plays a part in the effects of

The large, irregular mounds in the foreground and middle distance form the terminal moraines of the Cameron Glacier on the east side of the Southern Alps of New Zealand. The readily frost shattered greywacke rocks and the relatively dry climate on the rain-shadow side of the Southern Alps provide much debris and not too active meltwater streams. The sharp-crested lateral moraines of the shrunken present glacier are visible in the centre of the picture. Frost-shattered peaks and small corrie glaciers can be seen in the background.

glaciation in the uplands. The influence of meltwater increases towards the snout of the glacier and is effective beyond it, both in erosion and deposition. Beneath the ice it will be most effective in temperate glaciers, where it can penetrate to the bed. Subglacial meltwater can flow uphill under hydrostatic pressure. By this means subglacial meltwater streams can create the up-and-down profiles typical of some meltwater channels, such as those in north Northumberland. Meltwater channels are typically deep and flat-bottomed. On a smaller scale, meltwaters can create moulins, or glacier mills, that penetrate into the sides or floors of glacial valleys.

Temperate glaciers do not form such efficient dams to impound glacial lakes as cold glaciers. Features characteristic of the cold ice masses are marginal drainage channels, such as those cut around the shrinking Barnes ice cap in Baffin Island. This ice cap also impounds large marginal lakes, such as Generator Lake, which would drain to the west coast in the absence of the ice sheet across its former valley. The lake now drains to the east coast. In this way important drainage changes can be initiated; there are good examples of this in Pembrokeshire in South Wales.

Glacial meltwater carries a heavy load of coarse and fine debris from the ice, incidentally indicating how efficient glaciers must be as erosional agents. This material can be laid down in tunnels under stagnant or slowly moving ice to form long, sinuous ridges, known as *eskers*. In the uplands, however, the ice is often not slow moving enough to develop these forms, and *valley trains* are more common. These are gravel spreads over which the meltwater flows in braided channels when it leaves the ice. The valley trains have a flat cross profile with several shallow, but fast-flowing, streams, while the long profile is relatively steep, owing to the large and coarse load carried by the water.

Lakes can form around the margins of glaciers. They can be dammed in tributary valleys, and at the snout of glaciers, where they are often impounded behind moraines. Occasionally they occur beneath the ice, as under Vatnajökull in Iceland. Here, the subglacial lake of Grimsvotn is fed by meltwater, part of which is derived from geothermal melting. This water accumulates beneath the ice until it reaches a critical depth at which it causes the ice above to float. The water then escapes in a catastrophic flood, called a glacier-burst or, in Icelandic, a jökulhlaup. This flood occurs fairly regularly at about ten year intervals, although as the ice sheet is thinning the floods now occur rather more often. They are, however, smaller in volume as less water is needed to float the thinner ice and allow the water to escape suddenly. The jökulhlaup flood the whole of the outwash area at the glacier snout, and until very recently they have prevented the building of a road and bridge across the outwash plain, which is called a sandur in Iceland.

Glaciation in the lowlands

The effects of glaciation in the lowlands are typically the result of the action of ice sheets rather than glaciers. The lowlands are predominantly the area where the glacially derived material is deposited, although in some special situations the processes can be erosional. The deposits often hide the previous erosional effects under a thick blanket of till. The processes of deposition will be considered first, then the characteristics of the deposits, and finally their form and distribution will be mentioned. Some examples of lowland erosion in Britain will be mentioned in the next chapter.

Processes of glacial deposition

The characteristics of till give some clues concerning the processes of sedimentation. One process is by lodgement from beneath the ice. Figure 15 (page 31) shows that when the effective normal pressure exceeds a critical value for any given ice velocity, abrasion will cease and lodgement take place. Till is often lodged against protruberances on the bed and this is called *lodgement till*. A permeable bed will help the plastering-on process, whereby lodgement till forms. The process also involves shearing of the lowest debris-laden layers of ice, whereby fabrics are imparted to the growing deposit, which often takes place under advancing ice. The debris-rich lower layers of ice are often over-ridden by the more plastic cleaner ice above.

The lodgement process is controlled by many variables, including the temperature of the ice, the effective normal pressure, which in turn depends partly on the permeability of the glacier bed, and the bed roughness. A large flow law exponent, n, also causes more ready lodgement. The particle shape also affects lodgement, with plate-shaped particles lodging more readily than spheroidal ones. The process of shearing helps to align the particles in the till, often giving a preferred orientation to the long axes of the particles. The rate of subglacial lodgement has been observed by G.S. Boulton to be about 7 mm/day in a glacier in Svalbard.

Till can be released into cavities under the ice, where it is referred to as *slumped till*. The process is more efficient under temperate glaciers, as material can be released by the regelation process in this situation.

Another process producing glacial deposits is the melt-out process. This takes place mainly supraglacially, although it can occur both englacially and subglacially. *Melt-out tills* accumulate most prolifically where a large amount of debris has been entrained through regelation processes into the sole of the ice, and then carried up into the upper part of the snout along thrust planes. The material is deposited as the ice melts and is then let down onto the ground, often retaining a modified form of the original fabric of the till. When material melts on the surface of the ice and then flows down the ice surface it is called *flow till*. The tills are less consolidated than lodgement till.

The snout of an actively advancing glacier may bulldoze the material at its margin to form conspicuous ridges. The material may include previously deposited material of non-glacial origin as well as earlier glacial deposits. The process can produce features in which the original sedimentary characteristics are distorted. Low-angle faulting can occur when the material was previously frozen at depth. An example occurs in the coastal cliffs of north-west Wales at Dinas Dynlle.

Character of deposits

Material deposited directly by the ice is characteristically poorly sorted and unstratified, often consisting of large boulders in a clayey or sandy matrix, thus being a 'biomodal' deposit. The term 'till' is now normally used instead of boulder clay, which only strictly applies to deposits consisting of boulders set in a clayey matrix. The stones in till are often striated and sub-angular, due to the operation of abrasion and plucking. The elongated stones within the till are often arranged with a preferred orientation which normally indicates the direction of ice flow. The pattern of orientation of stones in the till is called the *till fabric*.

Ice is one of the most powerful transporting agents, and glacial deposits can include very large boulders. These are called *erratics* when they have been carried beyond their original geological outcrop. The Shap granite boulders that were carried up and over Stainmore to the east coast of England are good examples. Rocks have also been carried from Norway to the coast of north-east England. Some very large erratics have been recorded, such as one of 18 150 tonnes, measuring 24.5 m by 12 m by 9 m, in south-west Alberta. A Carboniferous limestone erratic in Anglesey measures 90 m in length, and one of Lincolnshire oolite at Melton Mowbray measures 275 m by 90 m. The largest erratic block of Germany measures 4 km by 2 km by 120 m and is composed of Tertiary and Cretaceous sands and clays which were frozen when they were moved and deposited.

Till fabrics have been studied to establish former ice flow direc-

The dark, erratic blocks, which are of pre-Carboniferous age, have been carried upwards by the ice from their outcrop a few miles away, and left on the light, Carboniferous Limestone at Norber Brow, near Austwick, North Yorkshire. The pedestal on which the large erratic block rests indicates the amount of limestone solution since the ice retreated.

tions. The processes of deposition, however are such that the preferred orientation is not always parallel to the ice flow direction. Some fabrics show both a parallel and transverse pattern, the latter being especially pronounced when the ice was constrained to flow up hill or into a narrowing trough. Tills also show variations of fabric upwards through the deposit, indicating changes of ice flow or process of deposition over time. In lodgement till, most fabrics are probably imposed as the till is laid down. In melt-out and flow tills, the fabric may be modified by the final depositional process. It often maintains an element associated with the movement of the ice that carried the material to its final position before melting.

Till fabrics can also be altered by later ice advances. The re-orientaton of the fabrics in tills of Alberta seems to be due to later shearing along closely spaced thrust planes. Detailed studies of different shaped stones has shown that shape also affects the final pattern. Fabrics tend to be stronger in lodgement till, which is also often over-consolidated due to the ice pressure when it was deposited. Melt-out tills, on the other hand, have weaker fabrics due to subsequent movement and the deposits are less consolidated. The dip of the particles as well as their orientation often provides useful information, and may be associated with the angle of thrust planes within the marginal ice.

Morphology of glacial deposits

There are so many different forms of glacial deposits that a table may help to clarify the nomenclature and types of features. Table 4 lists some of these features. They are divided first into linear and non-linear features, and the first group is further subdivided into features parallel to the ice flow and those transverse to it. The subdivisions within these groups are related to the processes of deposition. Forms laid down subglacially often show evidence of the ice flow. They can be either parallel to the flow, as drumlins, or transverse to it, as Rogen moraines.

The change in orientation relative to ice flow has been associated with zones of deposition between the ice source and its margin. D.E. Sugden and B.S. John, for example, suggest six zones as shown in figure 19. The innermost zone nearest to the ice source is one of fluted ground moraine, followed by one of drumlinised ridges, beyond which is a belt of true drumlins. These are followed by a transitional zone with drumlins, eskers and some Rogen moraines. Next lies a zone of unorientated disintegration features, followed by the last zone of end moraines.

The essential feature of this and other similar suggestions is that a zone of streamlined features occurs closest to the ice source, where the ice is more active, while the outer zone nearer the ice edge is characterised by features associated with stagnant ice.

The change from drumlins to Rogen moraines may be associated with the change from extending flow to compressing flow, as suggested in figure 20. The downward trending slip planes of the extending flow allow streamlining to mould the drumlins into their typically elongated form, with a steep upglacier stoss end and long tapering tail. The features of the transitional form are more amorphous, but as the stress increases in the down-glacier direction till is lodged especially in the zones of lower stress where the shear planes carry ice upwards and debris with it. The Rogen moraines are formed parallel to the elongation of the slip planes and transverse to the ice flow. The Rogen moraines can be over 1 km long, 10 to 30 m high, and spaced 100 to 300 m apart. The preferred orientation of the stones is transverse to the ridge crest and parallel to the ice flow. This pattern could be explained by the movement of the material up along thrust planes. The formation of both these features is associated with active ice.

Drumlins are one of the most easily recognised feature of glacial

Linear features		Non-linear features
Parallel to flow	Transverse to flow	Unorientated
Subglacial forms with streamlining	*Subglacial forms*	*Subglacial forms*
a. fluted ground moraine	a. Rogen or ribbed moraine	a. low relief ground moraine
b. drumlins and drumlinoid forms	b. de Geer or washboard moraine	b. hummocky ground moraine
c. crag and tail	c. Kalixpinnmo hills	
	d. subglacial thrust moraines	
	e. sublacustrine moraines	
Ice-pressed forms	*Ice-pressed forms*	*Ice-pressed forms*
longitudinal squeezed ridges	minor transverse squeezed ridges and corrugated moraine	random or rectilinear squeezed ridges
Ice-marginal forms	*Ice front forms*	*Ice surface forms*
lateral and medial moraines, some interlobate and kame moraines	a. end moraines	a. disintegration moraine controlled
	b. push moraines	b. disintegration moraine uncontrolled
	c. ice thrust/shear moraines	
	d. some kame and delta moraines	

Table 4 Features of glacial deposition

45

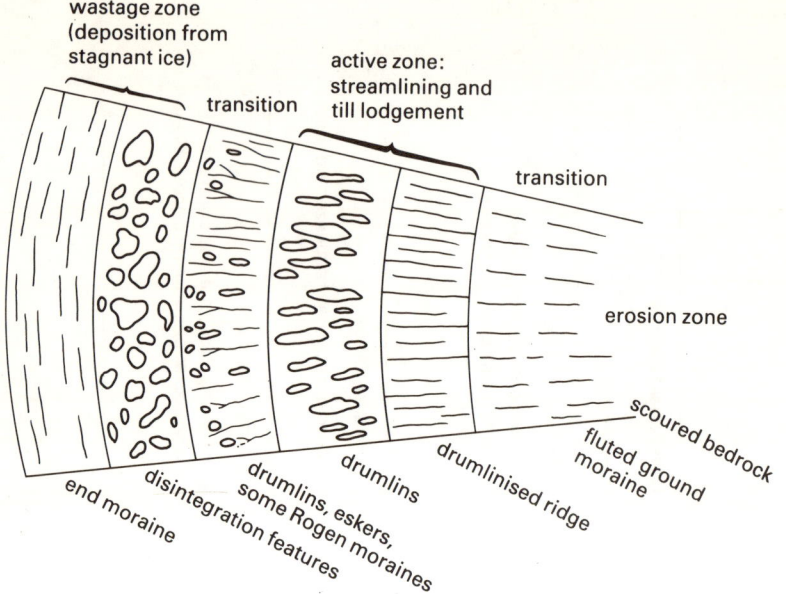

Figure 19 The six theoretical zones that occur between actively eroding ice and its outer margin (after D.E. Sugden and B.S. John, 1976)

deposition, but their formation is not easy to explain and many theories have been put forward. One theory to explain their occurrence in certain well defined belts has been proposed by I.J. Smalley and D.J. Unwin. They consider the *dilatancy* of till consisting of larger particles in a finer matrix. The mixed material is 'dilatant' in that it will expand under certain conditions. It lies between the moving ice and the bedrock and within the lowest layers of ice. If the flow of the lowest, till-laden ice is obstructed by an obstacle on the bed, it dilates, or expands, and the material becomes more resistant to shear under these conditions of dilation. The cleaner ice above then flows round the resistant lower, till-laden layer, shaping it into a drumlin with streamlined form.

Dilation continues as the stress increases until a critical value is reached: then the strength of the till-laden ice suddenly decreases and it can all be swept away by the ice. Under these conditions of high stress, therefore, no drumlins can form. If the stress is greater than the upper critical value no deposition can occur, but if it is less than the lower critical stress no deformation is possible. According to the theory, therefore, drumlins form where the stress is between the two critical values, that is between the minimum stress necessary to initiate dilatance and the minimum stress necessary to

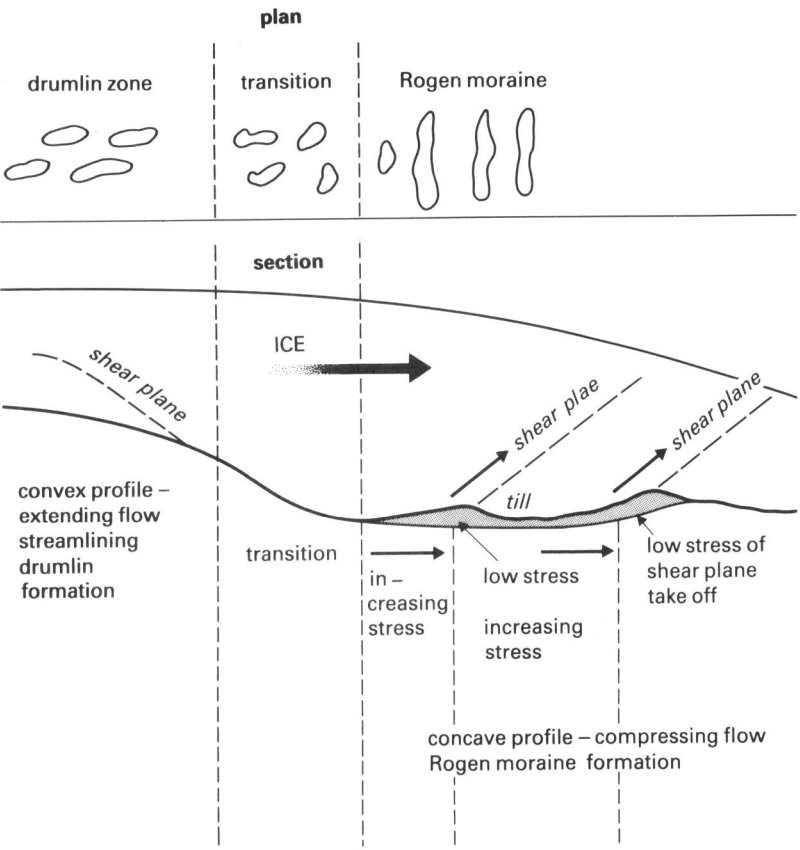

plan

drumlin zone | transition | Rogen moraine

section

ICE

shear plane

shear plae

shear plane

convex profile –
extending flow
streamlining
drumlin
formation

transition

till

in –
creasing
stress

low stress

increasing
stress

low stress of
shear plane
take off

concave profile – compressing flow
Rogen moraine formation

Figure 20 Diagrams to illustrate the possible transition between drumlin formation and Rogen moraine formation, shown in plan and profile

maintain it. Thus no drumlins can form where the ice is very thick or moving very fast, nor can they form where it is too thin. Thus drumlins tend to occur in well defined 'drumlin fields', at a point where the stresses reach the critical range.

More recent theories of drumlin formation suggest that it may depend on the relationship $S = C + (P - P_w) \tan \phi$, where S is the shear strength of the drumlin-forming till, C is the cohesion of the till, P is the ice pressure and P_w is the water pressure, and ϕ is the angle of internal friction. Any one of the variables could affect drumlin formation. A decrease of the pore water pressure, P_w, could result in an increase of till strength, enabling a drumlin to form around a stronger nucleus of till. This could occur at a permeable patch in the glacier bed. Local freezing could cause an increase in C, the cohesion of the till, and this again could lead to drumlin formation. The strength of the subglacial till is likely to increase up-glacier, and at some point the drag force imposed by the moving ice could cause bed deformation, suggesting that P, the ice pressure, is a critical factor. There is a critical zone of drumlin formation. On the up-glacier side it is bounded by a stress zone which is too high for the necessary obstacles to form in initiate drumlins. On the down-glacier side it is bounded by a zone in which the shear stress is not sufficient to deform the bed of till. In the critical zone, part of the bed must be deformable and part not. The part that is not deformable is shaped into drumlins and the rest is not. Till deformation begins as the stress level rises above the critical level. Again, drumlins will form in the zone where the critical conditions apply, sometimes in large numbers.

One of the largest drumlin fields occurs to the south of Lake Erie in New York state. Another large drumlin field occurs in north-east Ireland, again forming a well defined belt. Many drumlins form a ring around the Lake District, and these show by their elongation the importance of the constancy and speed of ice flow. They are elongated where the combined Lake District and Scottish ice was flowing west to the Irish Sea north of the Lake District, but are nearly circular where the ice was dividing the flow through the Tyne gap and over Stainmore out of the Eden valley near Appleby.

The streamlined form of drumlins can be appreciated by comparing their outline to a rose curve, which is an ideal streamlined shape. This curve is given by $p = L \cos k\theta$, and $k = L^2 \pi / 4A$, where L is the length of the drumlin and A is its area, k is related to the elongation of the drumlin and correlates closely with L/W, where W is the width of the drumlin. A common value for k is 3, although it can range between 1.5 and 10.

While drumlins are formed by active ice, a large number of glacial deposits show that they were formed when the ice was stagnant and decaying. Some of these features show alignment inherited from when the ice was active, others show no systematic pattern. Landforms referred to as *kame* and *kettle relief* consist of irregular gravelly hills and hollows, many containing shallow lakes. These occur where buried lumps of ice subsequently melt. The relief of such areas often becomes inverted as the buried ice melts.

Prairie mounds were initiated as hollows in the ice which filled with gravelly debris from meltwater drainage. Then the surrounding, cleaner ice melted, leaving a small mound, usually circular in form. They are common in Alberta in Canada. The mounds are usually about 100 m wide and 4.5 m high.

Eskers also form most readily in stagnant, temperate ice through which streams can drain, and in which stratified sandy or gravelly deposits are laid down. The surrounding ice melts to leave a long, sinuous ridge. The esker ridges sometimes feed into *delta-moraines* deposited in proglacial lakes. This is especially common when the ice is retreating down slope. There are good examples in Ireland in the Boyne valley near Dublin. Delta-moraines are formed of stratified sand and gravel.

Not all glacial deposits laid down in water are stratified. The de Geer or cross-valley moraines of Baffin Island are probably deposited annually at the margin of a quickly retreating ice sheet that ends in a water body. The till may be squeezed up in a water-soaked state to form a linear ridge. The material has a strong fabric on the side nearest the ice (*proximal*), and a random one on the opposite side (*distal*). The Baffin Island moraines of this type were formed under water, but somewhat similar ones have been observed to form subaerially in Iceland. Similar forms are also found in Sweden.

Not all deposits are associated with temperate ice. One type of cold ice moraine has been called a Thule–Baffin moraine from its occurrence in Greenland and Baffin Island. It was thought that the material was brought to the surface along thrust planes near the ice margin, and this may in fact occur. The moraines, which have ice cores when first formed, could also have originated by the inclusion of regelation debris into the ice far from the margin where the ice could reach pressure melting point. Near the margin it is frozen to its bed. The more active ice from up-glacier could over-ride the stagnant marginal ice and carry the debris upwards. It then emerges on the ice surface near the margin of the ice cap to form a con-spicuous moraine. Once the ice core has melted the moraines are very low and amorphous in form. Wind-drifted snow and buckling of the marginal ice may also help to form these moraines.

Even cold glaciers can produce much meltwater as they retreat, and large deltas and dead ice complexes exist in western Baffin Island. The area to the south of Vatnajökull ice cap in Iceland also illustrates the importance of meltwater, in this instance around a temperate ice mass. The meltwater has deposited an extensive outwash plain, the Icelandic *sandur*, which consists of large spreads of stratified material, becoming finer towards its margin, where it is sandy. Similar large outwash plains occur in Denmark and north Germany around the former margin of the Scandinavian ice sheet. Tunnel valleys are associated with these outwash spreads. They were formed subglacially by strong meltwater streams, the braided nature of which formed wide, fairly shallow valleys. The Urstromtäler of Germany and Poland are similar broad valleys, but they were formed along the margin of the retreating ice sheet, when the ice was retreating towards the Baltic and holding up the normal drainage northwards, and thus flow mainly east to west.

Glaciation of the British Isles

The ice age has had a profound effect on the British Isles, even though it is unlikely that the whole country was ever covered by ice. There are records of at least three major ice advances in Britain, with a number of minor oscillations and late glacial re-advances. The chronology of the ice advances will first be mentioned briefly, then the ice sources and flow patterns will be considered. The main features resulting from the older advances will then be examined, followed by an account of the effects of the later advances until the final disappearance of the ice from Britain about 10 000 years ago.

Chronology of the ice advances

Fourteen stages have been recognised in the British Pleistocene, but ice only developed in the country during the last three cold phases. The earliest ice advance took place in the Anglian glaciation, although there is evidence of climatic cooling in some of the earlier stages. The Anglian glacial deposits overlie the members of the Cromer Forest Bed Series, which consists of tills with layers of sand. Above these deposits is the Norwich Brick Earth which is overlain by the main glacial tills of the Lowestoft period. The Corton sands that underlie the Lowestoft tills appear to belong to a cold sea and are now thought to have only interstadial status. The glaciation to which all these deposits belong has been called the Anglian. This period was, therefore, a long and complex one with ice retreats intervening between major advances.

The succeeding interglacial is called the Hoxnian, after Hoxne in Suffolk, where a deep hollow on the till surface accumulated a sequence of deposits. These have yielded a long pollen sequence that indicates true interglacial conditions during this period. In some places there is evidence of both warming and cooling phases in this interglacial, although the latter part is missing from Hoxne itself.

The next major glaciation is called the Wolstonian and its type areas are mainly in the Midlands, the name coming from Wolston in Warwickshire. There are interglacial deposits in the Midlands that correlate with the Hoxnian of East Anglia, one site is at Nechells. In the Midlands the glacial deposits start with sands and gravels, which are the outwash of the advancing ice. These are overlain by till, which is itself capped with further gravels of the retreat phase.

The succeeding interglacial is called the Ipswichian. It was followed by the last major ice advance of the Devensian, which reached its maximum about 20 000 years ago. Ice was not so extensive at this stage. It did, however, reach into the north Midlands in the west, to near York and as far as north Norfolk in the east. The late Devensian includes pollen zones I, II and III, the first and third being cold periods, while zone II is the milder Allerød period. The final stage is called Flandrian, and it started 10 000 years ago, when the present interglacial started. It was a period of rapid sea level rise.

Ice sources and ice flow in Britain

Many of the highland areas of Britain supported their own ice caps, which at the maximum of each ice advance joined to form one continuous ice sheet. The British ice sheet also linked with the Scandinavian one over the North Sea basin. A model of glacial erosion and deposition over Britain has been proposed by G.S. Boulton and others. It reconstructs the conditions in Britain at the maximum of the Devensian ice advance. At this time the summit of the ice sheet is estimated to have been about 1800 m above sea level and velocities in the marginal zone would have ranged between 150 and 500 m/year. The ice would have been cold at its base in the central area, but temperate towards the margin. The ice sheet would have needed about 15 000 years to grow to its maximum. It is suggested that the lobe of ice that extended as far as north Norfolk on the east coast was the result of a surge (see p. 26).

The input for the model includes the maximum extent of the Devensian ice, the flow lines, which were determined from the distribution of erratics, and the accumulation. The precipitation was estimated to have been 1800 mm in south-west Ireland, decreasing to 700 mm in north-east Scotland.

The main centres of outflow of ice were the western Scottish Highlands and south-west Scotland. An ice-parting extended south across the Lake District and into north-west Wales, both of which had their own centres. There were also separate centres over Ireland. The ice speeds increased towards the margin of the ice sheet, and would have been relatively low in the central areas of dispersal. The ice sheet would have had high ablation and accumulation rates, and hence have been an active ice mass. The high ablation rate was partly due to calving into the Atlantic Ocean. The total ice sheet volume is estimated as 346 000 km^3, or equivalent to a world-wide sea level fall of 0.96 m. Figures 21 and 22 illustrate some aspects of the model.

Although the marginal zone would have been temperate ice probably, the extreme edge of the ice sheet could have been frozen to its bed, especially as it advanced over permanently frozen

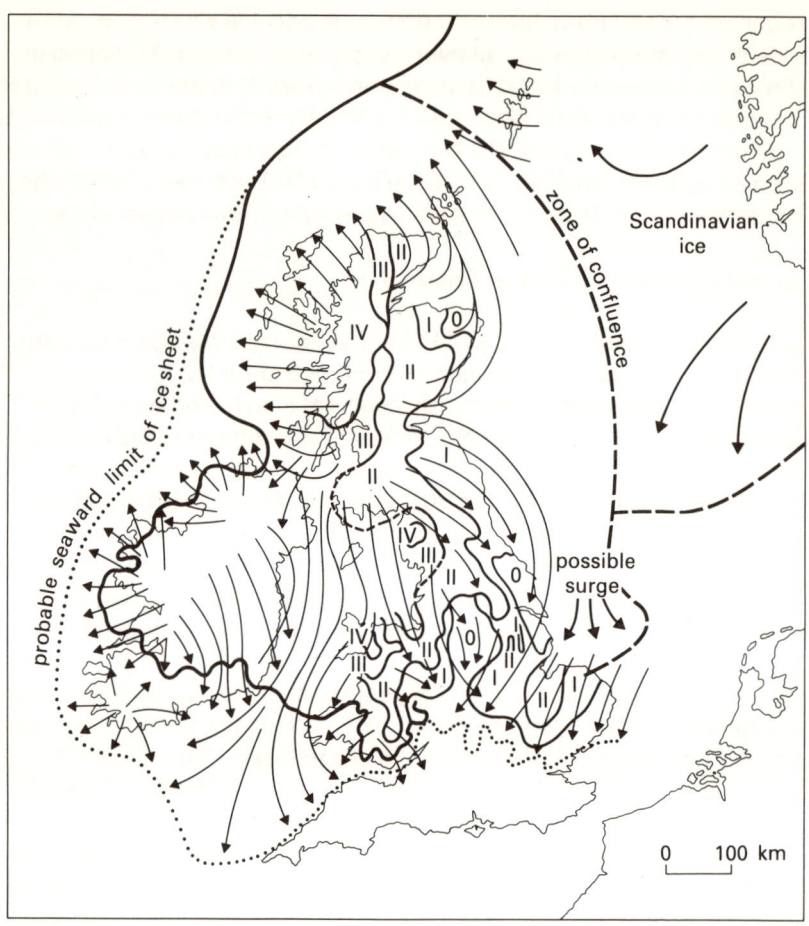

Figure 21 Ice limits and ice flow patterns over the British Isles. The Roman numerals indicate zones of increasing erosional intensity from 0 to IV (after G.S. Bolton *et al*, 1977)

ground. The ice sheet would have been erosional in the central areas of dispersal in the highlands. There would also have been a zone of strong erosion towards the margin, where the ice was flowing fast and was at the pressure melting point.

The volume of the chalky till of eastern England suggests that this locally derived material must have been eroded by an actively eroding ice mass. This is particularly noticeable in the chalk area of East Anglia, where the chalk scarp has been almost entirely eroded away, whereas to the south in the Chilterns it has a height of 250 m. The scarp crest further north has probably been lowered by about 100 m. This massive marginal erosion would be assisted by the

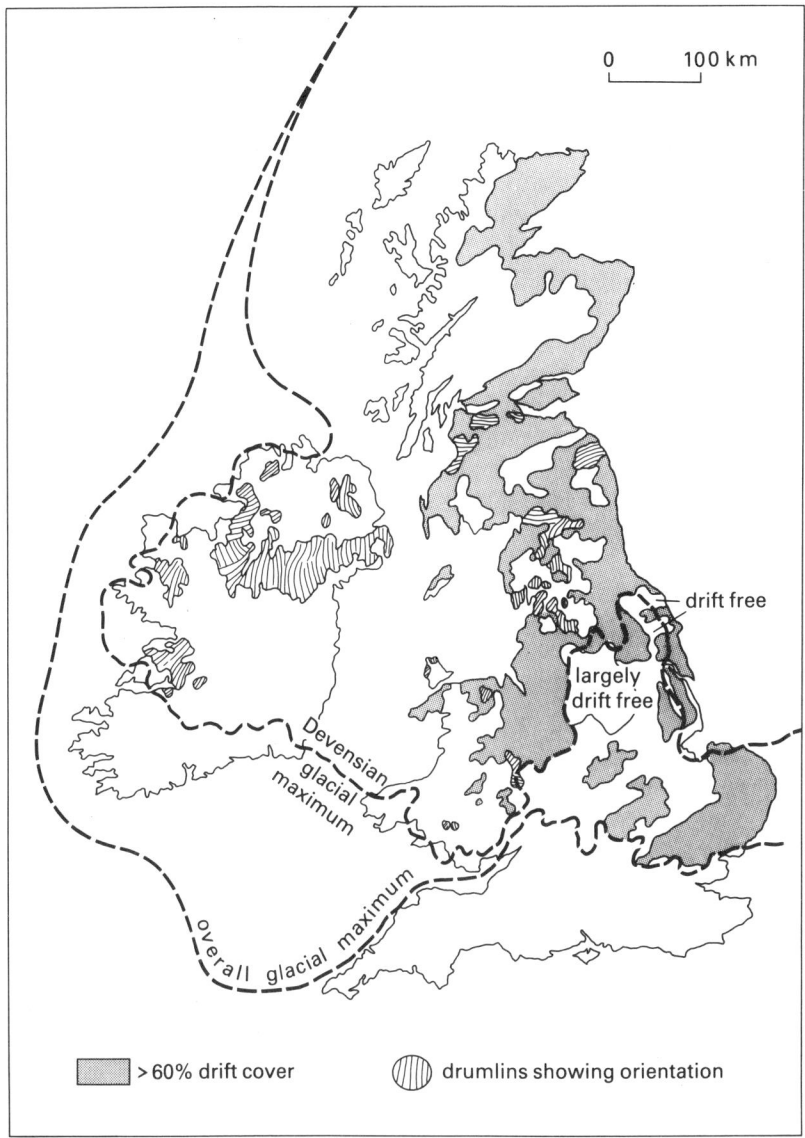

Figure 22 Limits of glaciation and distribution of drumlins and areas with more than 60 per cent drift cover over the British Isles (after Bolton _et al_, 1977)

rapid flow of the ice and also by the permeability of the bed, which would allow water to drain away and so increase the effective pressure, as this is the ice pressure minus the water pressure. Erosion in the highland zone would probably have been less at the maximum of the glaciation, as the ice would have been frozen to its bed and moving slowly. The large amount of erosion in the uplands probably took place under different conditions, as the ice was advancing or retreating.

The ice sheet model also provides some information concerning the pattern of deposition. It is suggested that the drumlin belts occur where the till deposits were thick, and where the basal sliding velocity was less than 50 to 100 m/year. The slower flow occurred in the zone nearer the centres than the fast-flowing, erosional marginal zone. Hence the main drumlin fields lie further north and nearer the ice centres than the marginal erosion zones.

Another zone of deposition is that characterised by kamiform relief due to the melt-out of thick englacial deposits from stagnant ice. This zone occurs mainly in the marginal areas. The material is probably derived from areas of net freezing at the base of the ice sheet, as it moved over the permafrost area and was cooled as a result. This applies, for example, in the Cheshire and Shropshire area.

The earlier glaciations in Britain

The evidence for the earlier glaciations is most clearly seen in central and eastern England. The ice of the Anglian glaciation reached furthest south. The early phase of the Cromer till left deposits on the north Norfolk coast, while the overlying Lowestoft till has been identified from Lincolnshire through East Anglia to the Thames estuary, where it formed the Hornchurch lobe. It was this ice lobe that probably diverted the Thames into its present valley from a more northerly course, and initiated the Lower Lea valley. These diversions most likely took place during the retreat phase of the Anglian ice advance.

In the Midlands there is also evidence of ice at this stage, where a lower layer of till has been found. In the Oxford area it is in the form of plateau drift, while in the Leicestershire–Warwickshire area it is called the Bubbenhall clay. The deposits of this advance are separated from those of the following one in the Midlands by inter- glacial deposits found at Nechells and Quinton near Birmingham, which are the equivalent of the Hoxnian of East Anglia. The Boyn Hill and Swanscombe terraces of the Thames also date from this warm interglacial period.

The Wolstonian glaciation produced profound changes in the drainage pattern of the Midlands, but does not seem to have been so effective in East Anglia, where it apparently left few deposits. The ice advancing into the Midlands laid down the Baginton–Lillington gravels in the north-trending valley, in which the advancing ice then dammed up a large lake. This glacial lake has been called Lake Harrison, and the clays laid down in it include the Bosworth clays. The south-western outlet of the lake was dammed by the Welsh ice, so that it overflowed over the Jurassic scarp to the south-east, as indicated in figure 23. The ice later advanced over the site of the lake as far as Moreton-in-the-Marsh, obliterating the lake and depositing till over the lake clays. When the ice retreated the present lower Warwickshire Avon valley was initiated.

Later the Severn was diverted into this drainage system through the Iron Bridge col. This col formed the outlet of a lake dammed between the high ground and the retreating ice to the west. These events probably took place within the Devensian, when the ice front was less advanced than during the Wolstonian glaciation. Ice did, however, extend as far as Wolverhampton in the west Midlands in the Devensian glaciation. The Severn and Avon terraces, therefore, span two glacial epochs and the intervening interglacial.

The Devensian glaciation

Devensian ice covered most of highland Britain, including most of Ireland apart from some of the south and south-west. A large lobe spread down the Irish Sea to the south-east tip of Ireland and to St. David's Head in south-west Wales. A lobe of Welsh ice extended almost to the south coast of Wales in the Swansea–Cardiff area. The ice front then swung north, only reaching as far as Wolverhampton and covering the Shropshire–Cheshire lowlands. Most of the southern Pennines were not ice covered at this stage, although the Askrigg Block and northern Pennines had an extensive ice cover. The ice probably did not cover all the hill tops towards the margins of the Dales. A lobe of ice extended down the Vale of York as far as the Eskrick moraine, but the North Yorkshire Moors were not covered. Ice impinged into Holderness, across the Humber estuary and eastern Lincolnshire and as far south as the north Norfolk coast. In the area to the south of the glacial limit the ground became permanently frozen and permafrost features are widespread, including fossil ice wedges and patterned ground.

As the ice advanced it created many of the well known features of glacial erosion. When it was still flowing actively it streamlined the till into the drumlin fields in those areas where conditions were suitable.

During the retreat of the ice a great deal of meltwater was available and large glacial lakes were widespread. The ice that moved down the Yorkshire coast impounded the waters of the Derwent

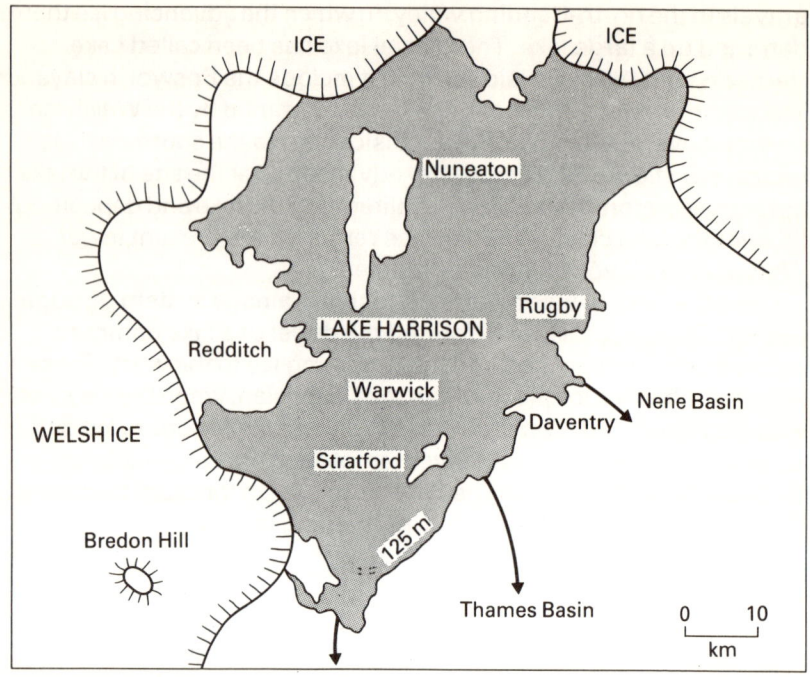

Figure 23 Glacial Lake Harrison at its maximum, showing ice dams and overflow routes

river to the north of Scarborough. This lake overflowed through the Forge Valley channel into a still larger lake held up by the ice in the Vale of Pickering. Water from Lake Pickering in turn overflowed through the Kirkham Abbey channel into an even bigger lake, called Lake Fenland. The meltwater eventually reached the sea through the southern North Sea basin and the Straits of Dover, then probably dry land.

A lake on the west side of the Tyne gap reversed the drainage of the South Tyne, diverting it to the east away from its former course to the west towards Carlisle and the Solway Firth. There were also complex drainage changes in north-east Wales.

One of the best known features associated with the ice-dammed lakes is the sequence of lake terraces which are called the parallel roads of Glen Roy in Scotland. The lakes were dammed up in three glens with lake levels at 355 m in Glen Gloy, 350 m in Glen Roy and 260 m in Glen Spean. The lake in Glen Roy was 16 km long and up to 200 m deep. The terraces occur at 325 m and 260 m, at the levels of the various lake outlets that were open at different times as the ice retreated. In places the 'road' is a kame terrace. A kame terrace forms where meltwater flows between the ice and the valley side, leaving a layer of deposits. A terrace forms as the ice melts. At its largest, the lake system amalgamated into one lake 35 km long, and it drained north-east to the Moray Firth.

The final decay of the ice sheet produced widespread fluvioglacial forms. The ice surface wasted downwards revealing first the hill tops, the ice being restricted to the valleys where it became sluggish and finally stagnant. Most of Scotland was ice-free before the Loch Lomond re-advance, which belongs to the late glacial period.

The late and post-glacial periods

A series of fifty-eight sites indicates a late-glacial interstadial in Scotland, with radio-carbon dates ranging between 12 500 BP (before the present) and 13 150 BP. Because the material analysed must be organic for radio-carbon dating, these sites show that Scotland was ice-free by 13 000 radio-carbon years ago, and there is similar evidence for a rapid climatic amelioration in Wales and England. The last major ice advance in Scotland produced the Loch Lomond re-advance. This is dated as zone III, lasting from about 10 800 to 10 300 BP. There are a number of dates for a warmer period before this final advance. They are referred to as zone II, the Allerød warm interval. The ice of the Loch Lomond advance was centred on the western Grampians and lobes extended as far as the southern end of Loch Lomond, and glaciers advanced down many of the highland glens. There was also a small ice cap on Mull and many other high areas had active cirque glaciers, including Skye, north-west Scotland and the Cairngorms.

Ice also occupied the cirques of the Lake District and Snowdonian mountains at this time, which was the last occasion that ice has existed in Britain. It seems likely that only long-lasting snow patches existed in the Pennines and other highland areas of Britain during this period. The higher hills of Ireland probably contained cirque glaciers, including those of Kerry and the Wicklow Mountains near Dublin.

During the maximum of the last major ice advance temperatures were probably 10–12° C lower than at present. The low temperatures lasted until about 16 000 years ago. A period of rapid amelioration took place between 10 000 years ago and about 7000 years ago. Since then there have been relatively minor oscillations. The coldest of these oscillations occurred between AD 1550 and 1850, a period usually referred to as the Little Ice Age. It was marked by ice advances to the historic maxima in Europe and elsewhere, but ice did not develop in Britain during this period.

Conclusion

The significance of glaciation

The ice age has affected the whole earth in many ways. Large areas in the higher latitudes have been directly affected by ice sheets and glaciers. These have left evidence of their passage over the landscape by characteristic deposition and erosion. Beyond the area actually covered by moving ice the landscape has been widely affected by permafrost. Periglacial processes have been very active around the lower latitude margins of the glaciated areas, and in some high latitude areas that were too dry for glaciers to form. The latter include eastern Siberia, parts of the Canadian Arctic and small areas of Antarctica.

Beyond the zones directly affected by glaciation or by permafrost, lie the lower latitude zones of Mediterranean, desert and tropical climates. These are zones of seasonal or spasmodic rain. Then there is the Equatorial belt of nearly continuous rain and constant high temperatures. These climatic zones have shifted with the growth and decay of the ice sheets. As a result, in many parts of the world in lower latitudes there have been wetter and drier periods. These have been either in phase or out of phase with the glacial periods, according to the local circumstances. The wetter phases are referred to as pluvials.

It was thought at one time that the glacials correlated with the pluvial phases, but recent work has cast doubt on this assumption, especially in Africa. Lake Chad, for instance, was much larger during the interglacial phases. In North America, on the other hand, the large pluvial lakes, including Lake Bonneville, of which the Great Salt Lake is a very small remnant, was at its largest during the glacial periods. The situation is complex. The drier phases have been identified through the occurrence of large areas of fossil dunes, which are now vegetated, in wetter areas.

Another important function of the wind during the glacial periods in the higher latitudes was the widespread distribution of silt-sized particles. The material was derived from the extensive glacial outwash deposits and now forms the thick, widespread deposits of loess. These cover large areas of China, central Asia, North America and eastern Europe.

Perhaps one of the most important effects of the glacial period was the rapid swings of sea level that are associated with the growth and decay of the ice sheets. There are many causes of sea level change. None, however, act so rapidly and to such a large degree as the world-wide *eustatic* sea level changes associated with the withdrawal of water from the oceans to form the ice caps and its subsequent return. The *isostatic* changes caused by the weight of the ice concentrated on the areas covered by the ice sheets is also significant. Smaller changes of weight occur over the oceans as water is withdrawn from or added to them. The water is spread much more thinly over the oceans than the ice on land, but both affect sea level. The changes of sea level are extremely complex and few areas experience the same fluctuations.

Sea level has been more than 100 m below its present height during the maximum period of ice cover, and possibly nearly as much as 160 m. During the Devensian glaciation sea level was as low as 100 to 135 m below the present level. The latest change of sea level has been the Flandrian transgression, which was a very rapid rise between about 15 000 years ago and about 5000 years ago. Eustatic sea level was about 60 m lower than present 10 000 years ago; during the last few millenia, however, it has been relatively static. The recent variations have been within a range of about 10 m. Sea level is still rising slowly in many places, as the ice masses continue to diminish in volume.

The rapid swings of sea level have had a marked effect in many ways. They have greatly affected the development of coastlines all over the world, and influenced the lower reaches of the rivers. They also influence the spread of plants, animals and men, as land bridges come and go.

The future

At present all the indications suggest that we are approaching the end of a warm interglacial. It is likely that in the fairly near geological future the earth will enter another glacial phase. During the Little Ice Age a few centuries ago the conditions were nearly suitable for the development of the Laurentide ice sheet. Snow patches extended over the high plateaux of Baffin Island and elsewhere in northern North America. Once a threshold has been passed ice sheets can develop very rapidly by positive feedback processes.

There is, however, also the possibility that human interference with the atmosphere could reverse the cooling trend. The main effect is caused by added carbon dioxide, which is derived from the burning of fossil fuels and the destruction of forests, especially in low latitudes. It has been argued that the greater the amount of carbon dioxide in the atmosphere, the warmer the temperatures are likely to be. The problem is, however, very complex. It is not known how much human interference has affected the natural operation of the ocean–atmosphere circulation system. This system controls the climate, and thus the generation and decay of ice sheets, and the many repercussions that glaciation produces.

Topics for discussion and exercises

1   Locate cirques on Ordnance Survey 1:50 000 or 1:25 000 maps of upland areas, such as Snowdonia, the Lake District and Scottish Highlands. Note their orientation, the height of their lips and back walls. Plot the pattern of orientation and discuss the relationship between form, height and position of the corries in association with aspect and exposure.

2   From an Ordnance Survey map of a glaciated upland area, draw cross and long profiles of a glaciated valley. To the cross section, fit a parabolic curve, $y = ax^b$, where y is the height above the valley floor and x the horizontal distance from the valley centre line. Note how close to 2 the exponent b is.

Discuss the nature of the longitudinal profile, noting the presence of lakes and rock steps. Locate examples of different types of glacial troughs on the map, noting examples of diffluence and transfluence and the effect on the drainage pattern. (Maps of the Scottish Highlands are suitable.)

3   On a large-scale map (1:25 000) of a drumlin field, measure the drumlin elongation and orientation. Consider these characteristics in relation to the ice source and flow direction.

4   Make a rough estimate of the activity index of ice masses in different parts of the world, using temperature and precipitation values obtained from data supplied by an atlas or other source. The higher these values, the higher the index will be. List the ice masses in order of activity.

5   Examine the effects of glaciation in your local area or an area that you have visited on field work. Assess the relative importance of glacial erosion and deposition. If you live in an area beyond the glacial limit, assess the *indirect* effects of glaciation.

6   Discuss the possible results on the geography of the British Isles of both an increase and a decrease of ice volumes on earth. Consider the factors that may influence future glacial fluctuations, including human interference.

7   Consider the conditions under which ice becomes an effective agent of erosion, and locate areas where these conditions are, or have been, fulfilled. What evidence do they contain to support your considerations?

Further reading

J.T. Andrews, *Glacial systems*, Duxbury, Massachusetts, 1975

C. Embleton and C.A.M. King, *Glacial geomorphology*, Arnold, London, 1975

R.F. Flint, *Glacial and Quaternary geology*, Wiley, New York, 1971

J. Imbrie and K.P. Imbrie, *Ice Ages*, Macmillan, London, 1979

B.S. John, *The Ice Age past and present*, Collins, London, 1977

W.S.B. Paterson, *The physics of glaciers*, 2nd edn, Pergamon, Oxford, 1981

F.W. Shotton (ed.), *British Quaternary studies*, Clarenden Press, Oxford, 1977

D.E. Sugden and B.S. John, *Glaciers and landscape*, Arnold, London, 1976